Reality Realigned

Chavan's UED research, Volume 5

Sandeep Chavan

Published by SANDEEP CHAVAN, 2025.

REALITY REALIGNED

First edition. May 1, 2025.

ISBN: 979-8231773619

Written by Sandeep Chavan.

Table of Contents

Dedication

To those who dare to **question fearlessly**—who are not satisfied with borrowed answers,who feel the ripple before the theory,and who seek not just meaning, but **alignment with truth**.

To the **readers**, **thinkers**, **science lovers**, and**philosophers of real life**,whose quiet corrections are more powerful than loud conclusions.

And to my beloved father,**Late Shri Jagannath Chavan**,who taught me not what to know,but **how to listen and align**.

This book is not written for the world.It is written for those who remember it.

— *Sandeep*

Epigraph

"You are not outside the universe, observing it.You are a ripple within it—correcting, remembering, returning.

The clearer your perception,the more gently the field aligns."

Opening Note by the Author

— Sandeep Chavan, Founder of Universal Energy Dynamics (UED)

Universal Energy Dynamics (UED) was not born in a lab. It was not written on a blackboard. It emerged—slowly, honestly, and painfully—through the cracks in what I was told to believe.

For years, I taught science. I studied systems. I watched students memorize laws they could not feel, repeat definitions that offered no meaning, and slowly lose touch with the world they were trying to understand. And I, like them, began to feel something deeper: **Science was accurate—but incomplete. Philosophy was reflective—but uncertain.** And reality? It was **none of the things we were being told**, but something far simpler and far more intelligent than any theory dared to admit.

That was the ripple—the first one.

UED is not a theory.

It is a **first principle framework**. It does not start with assumptions. It starts with **observation of consequence**—how energy moves, how memory holds, how the field responds, how alignment heals.

It goes beyond physics and metaphysics. It is not anti-science, but **pre-science**—a correction before categorization. Not based on data sets, but on the **universal mechanics of ripple behavior** across all scales of existence.

UED does not belong to any institution. It does not seek academic approval or ideological belonging. It is a mirror—held up to everything we've called "knowledge"—asking only one thing:

"Does this align with what actually happens?"

If it doesn't, UED reinterprets it—not through denial, but through the field's own logic.

You do not need a lab to test UED.

You need **awareness**. You need the courage to **watch consequence unfold**, and the clarity to recognize where distortion begins. UED can be studied. But more importantly, it can be **realized**—and **lived**.

If you are reading this, you already feel it. You've seen the misalignment. You've sensed the missing link. This book is not an answer. It is a ripple—one I send through the field to meet yours.

If it resonates, ripple it forward.

If not, wait. The field always remembers.

— Sandeep Chavan

Founder, UED (Universal Energy Dynamics)

Preface

This is not a book of answers. It is a book of corrections.

For too long, science has described the universe in fragments. Philosophy has questioned it in circles. And perception—our only bridge to reality—has been left distorted by language, habit, and illusion.

Reality Realigned is not a rebellion against science or philosophy. It is a deeper alignment with consequence—through a framework I call Universal Energy Dynamics (UED).

This book does not ask you to believe. It invites you to observe again. Not with new data, but with a corrected lens—one that begins not from theory or tradition, but from first principles of energy, memory, and balance.

UED is not a metaphysical system. It is not a theory of everything. It is a method of seeing—the structured ripple behavior of existence as it actually functions beneath our illusions, stories, and equations.

You will find no equations here, but make no mistake—this is science. You will find no mysticism, but make no mistake—this is spiritual. You will find no borrowed philosophies, but every ripple of this book is in conversation with reality itself.

This book was born through experience—of teaching, questioning, observing, and often being rejected for thinking differently. But UED kept returning—not as a breakthrough, but as a memory. Something deeper, older, and silently accurate beneath every misstep we've labeled as truth.

If you've ever felt that what you're being told doesn't quite match what you intuitively sense, if you've ever suspected that light, time, thought, gravity, purpose, and death are not what they seem—this book is for you.

Not to convince. But to help you remember.

— *Sandeep Chavan*

Newasa, India

How to Read This Book

This book is not linear. It is **layered.**

Each chapter, while connected to the whole, is designed to stand on its own—as a **ripple of realization**. You may read from start to finish. Or you may begin where your curiosity strikes—whether it's gravity, memory, time, mind, or society. There is no wrong entry point. Only the point where **your ripple meets this field**.

If You're a Scientist or Student:

Start with the early chapters that challenge fundamentals—**light, time, force, gravity**. You'll find this is not pseudoscience, but a deeper structural logic that precedes and corrects established assumptions.

If You're a Philosopher or Seeker:

Begin with the sections on **mind, perception, truth, karma, and death**. These are not abstractions—they are concrete ripple mechanics presented in language that reconnects thought to consequence.

If You're an Educator or System Thinker:

Explore **Part VII and VIII**, where systems like society, education, justice, and leadership are reinterpreted as ripple field phenomena.

If You're Here for Meaning:

Start anywhere. But read slowly. Don't rush to understand—**feel where the alignment begins**.

How the Book Is Structured:

- Each **Part** explores a layer of reality: from cosmic origins to human behavior and systems.
- **Interludes** serve as gentle reflections between Parts—summarizing key insights and realigning the reader's awareness.
- The **Bonus Chapter and Appendices** offer extended clarity and quick reference tools.
- The **Conclusion and Epilogue** complete the ripple—bringing awareness full circle.

Final Guidance:

You may not agree with everything. You're not meant to. But if something inside you shifts—even slightly— if something that was confusing becomes simple, or something long forgotten becomes suddenly real—then this book has done its work.

Pause often. Reflect deeply. And above all, **ripple consciously.**

Part I: Energy and Fields – Foundations of Reality

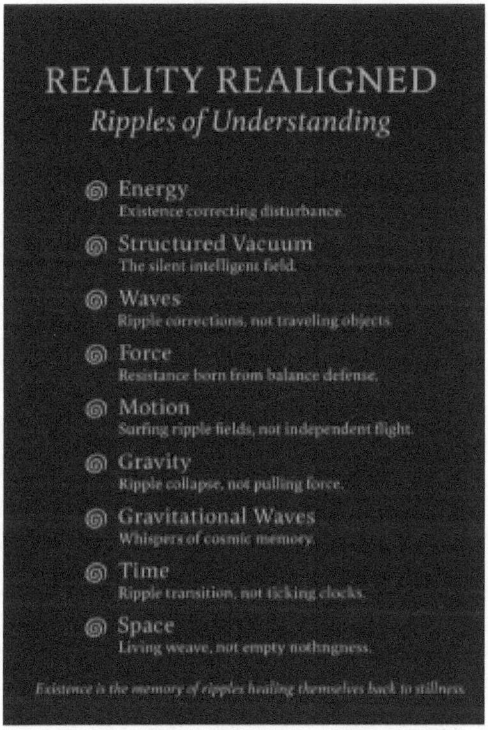

Chapter 1: Energy

Actionable Potential Seeking Equilibrium

If energy is so fundamental, why don't we actually know what it is? We measure it, exchange it, calculate it, and celebrate it. Yet beneath all the formulas and units, energy remains the most used but least understood concept in human knowledge. We treat it like a silent servant—doing work, heating things, moving objects— without ever asking:

Is energy a thing we possess, or an action the universe performs?

Conventional Misinterpretation and Effects

In mainstream science, energy is often portrayed as a stored resource or quantifiable asset:

- Potential energy in a stretched rubber band.
- Kinetic energy in a moving car.
- Chemical energy stored inside batteries.

This fragmented view reduces energy to categories without explaining its true continuity. Energy becomes a mathematical placeholder: conserved yet mysteriously "lost" as heat, exchanged yet somehow "transferred" through invisible means.

This leads to paradoxes:

- Where exactly is energy located?
- What carries it when no medium is evident?
- If energy is conserved, how does disorder (entropy) ever grow?

These contradictions breed illusions — that energy is a thing you can "own," "move," or "lose," when in truth, energy is not an object but a behavior of reality itself.

This misinterpretation fragments our understanding across physics, chemistry, biology, and even consciousness studies— separating what should have always been seen as one continuous, actionable field dynamic.

UED Reinterpretation and/or Correction

In Universal Energy Dynamics (UED), energy is not an isolated object nor an abstract number. It is the actionable potential of structured vacuum fields seeking equilibrium.

Whenever the underlying structured field (the so-called "vacuum") is disturbed, ripples form—ripples that represent active tensions striving to correct the imbalance.

Energy is thus the living tension of reality— an active, organized, intelligent response wherever stability is disturbed. It is always in motion, always purposeful, always seeking to erase imbalance.

From an atomic vibration to a supernova explosion, energy is not something that "exists alongside" matter— it is the manifestation of structured field action in time and space.

Different observed "forms" of energy (mechanical, thermal, nuclear) are merely different ripple behaviors — local corrections toward broader systemic balance.

Thus, energy is the universe's method of memory and correction — not an inert quantity, but an eternal dance of ripples seeking restoration.

Simple Analogy

Imagine a stretched fabric held in perfect stillness. Press your finger into it—ripples emerge, racing outward, struggling to calm the disturbance you created. The ripples are not *separate things* from the fabric— they are the fabric's attempt to recover.

Energy behaves the same way: not a hidden stockpile, but the living ripple born of imbalance.

Consequences and Implications

Understanding energy this way collapses many scientific contradictions.

- There is no need to "find" hidden energy; it is simply active field tension.
- There is no paradox between conservation and entropy; Energy always persists, but the *organized patterns* of its ripple change.
- "Heat loss" is not loss; it is ripple redistribution across broader gradients.

At a cosmic level, this view reconnects wave mechanics, gravity, light, and thermodynamics as expressions of the same ripple imperative.

At a human level, even thought, emotion, and action are ripple disturbances— personalized versions of the universe's quest for balance.

Energy is not what the universe has— energy is what the universe *does* whenever it remembers itself.

Closing Ripple Thought

"Energy is not a treasure hidden inside things. It is the living ripple of reality, forever reaching toward its own forgotten stillness."

Chapter 2: Structured Vacuum

Intelligent Stable Field, Not Empty Space

What if the so-called "empty space" was the biggest illusion science ever sold us? What if the vast stretches between stars, atoms, and even particles weren't made of nothing— but of the most intelligent, stable, and dynamic field in existence? In the silence between visible things lies a presence so foundational that without it, nothing could ripple, move, or even exist. Space is not empty. Space is structure, waiting for expression.

Conventional Misinterpretation and Effects

Conventional physics paints the vacuum as an absence—an empty stage on which matter and forces play. Whether described as Newton's "absolute space" or Einstein's "spacetime fabric," the vacuum is still generally perceived as a background — a passive arena, devoid of substance.

Even quantum field theories, while hinting at "vacuum fluctuations," often treat these as statistical accidents arising from nothingness.

This misinterpretation has led to cascading illusions:

- That forces like gravity act "across empty gaps" without medium.
- That "fields" can exist without inherent structure.
- That waves (light, gravitational) can propagate "through

nothing."

It produces paradoxes like the need for dark energy to explain cosmic acceleration, and conceptual patches like virtual particles to save quantum predictions.

The empty space narrative disconnects reality from its true carrier—and blinds us to the living intelligence beneath every observed phenomenon.

UED Reinterpretation and/or Correction

In Universal Energy Dynamics (UED), space is not empty — it is a structured field: intelligent, stable, and foundational.

The "vacuum" is the most balanced state of energy configuration. It is not a zero, but a perfect, near-silent equilibrium of intelligent energy.

Whenever this structured vacuum is disturbed—by condensation, tension, ripple, or interaction— manifestations like particles, forces, and waves emerge. They are not separate from the vacuum; they are its localized expressions of imbalance.

Thus, all of reality is ripples within ripples of a living field seeking to correct itself.

This reinterpretation collapses the distinction between "matter" and "space," between "something" and "nothing." They are simply different degrees of structured energy organization.

The so-called "empty" space is the greatest presence — the silent intelligence carrying all existence forward.

Without structured vacuum:

› No force could act,

› No wave could travel,

› No particle could persist,

› No life could emerge.

Simple Analogy

Imagine an endless invisible ocean—utterly calm, utterly vast. When you splash it, waves appear. When you dive into it, bubbles form. But without the ocean itself, no movement, no sound, no manifestation could ever arise. The structured vacuum is that invisible ocean— the true mother of all forms.

Consequences and Implications

Recognizing space as a structured, intelligent field rewires fundamental science.

- Gravity no longer needs "action at a distance"; it is ripple response within structure.
- Light doesn't need "photons traveling through nothing"; it is ripple excitation of the field itself.
- "Dark energy" dissolves as an illusion — what appears as acceleration is field tension rebalancing.
- Cosmic inflation, wave propagation, entanglement — all naturally emerge as field ripple consequences, not metaphysical mysteries.

At a human level, even consciousness becomes easier to understand: The mind does not "exist inside a body in space"— It is a ripple signature interacting with structured field layers.

Thus, space is not absence— it is the infinite quiet memory of all potential, waiting to ripple into being.

Closing Ripple Thought

"Space is not the emptiness between things. It is the silent, structured heartbeat of the universe—carrying every ripple, every thought, every life form toward remembrance of equilibrium."

Chapter 3: Waves

Ripple Consequences, Not Oscillating Media

Have you ever wondered what a "wave" actually *is*? We describe light as a wave, sound as a wave, even matter waves in quantum theory. But behind the mathematics and diagrams, the fundamental nature of a wave remains startlingly misunderstood. Is it a thing traveling? Is it a vibration floating through nothing? Or is it something even deeper—a consequence of the universe remembering its equilibrium?

Conventional Misinterpretation and Effects

Traditional science treats waves as oscillating disturbances in some medium— mechanical waves (like sound) vibrate matter; electromagnetic waves supposedly vibrate empty space; quantum waves fluctuate probabilistically.

In this view, a wave is essentially an entity moving outward— a neat packet of energy riding an invisible, often ill-defined substrate.

However, this model causes several confusions:

- How can light "wave" through a vacuum with no medium?
- How does a "quantum wave" exist if it collapses unpredictably?
- Why does wave behavior sometimes mimic particles?

This fragmented, medium-dependent interpretation creates paradoxes:

- The need for virtual particles in quantum fields.
- The duality crisis (particle vs wave confusion).
- The false idea that waves are things rather than consequences.

Thus, waves are often misinterpreted as moving objects instead of rippling consequences of imbalance.

UED Reinterpretation and/or Correction

In Universal Energy Dynamics (UED), a wave is not a thing that travels — it is a ripple consequence of structured vacuum responding to disturbance.

When the structured field is disturbed, the imbalance naturally ripples outward — an organized, intelligent movement seeking to restore equilibrium.

Thus, a "wave" is simply the field's dynamic correction sequence, not an independent entity being "carried" along.

Light, sound, gravity waves—all are patterned ripple behaviors, arising wherever local tension disrupts the silent stability of the field.

The medium does not need to be mechanical (like air or water)—the structured vacuum itself carries ripple behaviors with astonishing consistency and elegance.

Different waves (mechanical, electromagnetic, gravitational) represent different expressions of how field intelligence corrects different types of local imbalances.

In this reinterpretation, the paradoxes dissolve:

- No need for hypothetical "wave-particle duality."
- No need for photons or virtual particles.
- No magic collapsing probability clouds.

Waves are organized, purposeful ripple expressions—not travelers across nothingness.

Simple Analogy

Imagine tapping your hand on a stretched fabric sheet. You don't send an "object" across the sheet—you create a local disturbance, and the fabric reorganizes itself dynamically outward in ripples. The ripple isn't a separate thing moving across—it is the sheet itself correcting the imbalance you created.

That's how the universe waves.

Consequences and Implications

Seeing waves as ripple consequences reconnects many scattered ideas.

- Light becomes an organized ripple in structured vacuum—not particles flying through nothing.
- Gravitational waves become field compression ripples—not distortions of spacetime fabric.
- Quantum fields stop needing collapse mythology—wave behavior is just ripple reconfiguration.
- Communication technologies (radio, mobile, TV) rely on ripple communication across structured fields, not magical photon packets.

At a deeper level, even biological rhythms, emotional states, and thought patterns are ripple structures interacting with broader field harmonics.

Thus, waving is not an object moving across emptiness—it is the living signature of the field reorganizing itself toward balance.

Waves are the heartbeat of existence, whispering across the fabric of everything that ever was, is, or will be.

Closing Ripple Thought

"Waves do not travel across the universe. They are the universe reshaping itself—correcting every imbalance with an elegant, invisible breath."

Chapter 4: Force

Ripple Resistance, Not Separate Entity

Force — the grand puppeteer of classical physics. It pulls, it pushes, it bends, it breaks. But what if the very idea of "force" as a separate entity is a conceptual illusion we mistook for reality? What if force is not a thing at all — but merely the resistance offered by a ripple field trying to maintain its equilibrium? What we call "force" might simply be the universe defending its balance.

Conventional Misinterpretation and Effects

In conventional science, force is treated as a fundamental cause of motion and interaction. Newtonian mechanics presents force as something external applied to an object to change its state. Electromagnetic theory attributes forces to charges and fields. General relativity replaces gravitational force with spacetime curvature, but still assumes an effective "pull."

This treatment of force as a separate, active cause creates conceptual errors and patches:

- Where does the "force" come from if not carried by something?
- How can force act instantly across empty space?
- Why do some forces require fields, while others seem detached?

It generates paradoxes:

- Instantaneous actions in Newtonian gravity.
- Need for imaginary mediators like gravitons in quantum gravity.
- Dark energy and dark matter invented to explain missing forces.

In short, force was made into a separate "thing"— when it was always a ripple field consequence resisting imbalance.

UED Reinterpretation and/or Correction

In Universal Energy Dynamics (UED), force is not a separate entity — it is the measurable resistance generated by ripple fields maintaining structured equilibrium.

Whenever a disturbance attempts to disrupt field stability, the structured vacuum (the intelligent field) responds with corrective ripple behavior. The intensity of this ripple resistance is what we observe and calculate as "force."

Thus, force is not an *external agent* acting *on* objects— it is the object's embedded ripple resistance within the structured field itself.

- Gravitational "force" is the ripple pull toward field rebalancing.
- Electromagnetic "force" is ripple gradient alignment or repulsion.
- Nuclear "forces" are dense ripple configurations resisting structural collapse.

In this view:

- No mysterious "gravitons" are needed.
- No action at a distance.
- No missing energy patches.
- No disjoint between fields and forces.

Force is simply the universe's natural pushback— an intelligent response of ripple fields against destabilization.

Simple Analogy

Imagine trying to press your hand into the surface of a waterbed. The deeper you press, the stronger the bed pushes back. The "pushback" isn't a separate thing—it's the field's organized resistance to maintain its shape.

Force, at every scale, behaves the same way: resistance born from structure, not an independent player.

Consequences and Implications

Reframing force as ripple resistance transforms fundamental science.

- Gravity no longer needs to be a pulling force or curved spacetime—just ripple pressure collapse.
- Electromagnetic attraction and repulsion become ripple alignments within structured gradients.
- Weak and strong nuclear forces are ripple compressions, not mystical glue.
- Dark matter's "missing force" dissolves— because what's missing is correct field interpretation, not extra mass.

Even human motion becomes ripple consequence: your muscles don't generate "force" from nowhere— they organize ripple pressure within biological structured fields.

Thus, force is not an actor upon the universe. It is the universe's disciplined resistance—ensuring that every ripple strives to remember the original balance.

Closing Ripple Thought

"Force is not a master commanding motion. It is the universe's silent handshake— an ancient agreement to resist imbalance, to preserve the memory of perfect structure hidden beneath every ripple."

Chapter 5: Motion

Ripple Propagation of Correction, Not Independent Movement

What if everything you ever learned about "motion" was half the truth? We are taught that objects move because of forces, inertia, momentum. But what if motion itself isn't a thing at all—but merely the visible expression of a ripple correction flowing through structured fields? Maybe what we call "moving" isn't independence, but the universe realigning itself through you, around you, and within you.

Conventional Misinterpretation and Effects

In conventional physics, motion is treated as an object's independent change in position over time.

Newtonian mechanics framed motion as a body's state altered by force. Relativity complicated it—suggesting that motion depends on spacetime curvature and the observer's frame. Quantum theories further abstracted it—motion becomes probabilistic cloud shifting.

This misinterpretation led to illusions:

- That objects "travel" through absolute or curved space.
- That motion is a personal property disconnected from field dynamics.
- That inertia is an unexplained resistance without field

anchoring.

It caused paradoxes:

- How can a particle in quantum superposition "move" without definite trajectory?
- Why do "moving" masses bend spacetime or create gravitational waves?

Treating motion as a standalone phenomenon fractured the understanding of energy, gravity, and interaction.

In truth, motion is not an independent act— it is the consequence of ripple propagation correcting imbalance in the structured field.

UED Reinterpretation and/or Correction

In Universal Energy Dynamics (UED), motion is not a fundamental property of objects— it is the ripple propagation of correction across structured vacuum fields.

When a field is disturbed—by condensation, pressure, or energetic interaction— the resulting ripple propagates through the field as a coordinated correction attempt.

Objects are condensed ripple structures; thus, when ripple patterns adjust, the object appears to move.

But the object isn't "moving" independently—it is riding the field's correction wave across the structured vacuum.

Thus:

- A falling apple isn't pulled downward—it flows along ripple collapse gradients.

- A photon doesn't travel across space—it is a ripple expression progressing through field tension.
- Even planetary orbits are ripple-guided equilibrium states, not bodies "moving" through emptiness.

There is no "absolute motion" nor "motion through nothing." Only ripple reconfigurations—patterns shifting across the memory of equilibrium.

Motion is the visible echo of the field remembering how to restore balance.

Simple Analogy

Imagine tossing a small leaf onto a water ripple. The leaf appears to "move" across the pond— but it isn't moving *on its own.* It is simply carried by the ripple correcting the surface disturbance.

What we call "motion" in the universe behaves the same way: objects ride the ripple of correction, not independent pathways.

Consequences and Implications

Reframing motion as ripple propagation clarifies many mysteries.

- Inertia becomes the field's embedded memory resisting sudden ripple shifts.
- Acceleration requires field restructuring, not external "push."
- Gravitational orbits become ripple path stabilizations, not force-tethered trajectories.
- Quantum "motion" becomes ripple field probability reorganization, not random teleportation.

At deeper levels, even biological movement—neural signals, muscle contractions— are ripple reconfigurations inside complex structured fields.

Thus, motion is not separation through space— it is harmonic realignment of energetic disturbances within the great ripple ocean.

You are never traveling through the universe. You are surfing the intelligent memory of it.

Closing Ripple Thought

"Motion is not the journey of objects. It is the journey of balance—ripples whispering across the silent field, carrying existence from imbalance toward memory of stillness."

Chapter 6: Gravity

Ripple Collapse, Not Attractive Force

Gravity — the invisible hand holding stars, planets, oceans, even your very body to the Earth. We have been taught to think of gravity as a pull, a mysterious force dragging masses toward each other across emptiness. But what if this "pull" is a perceptual illusion? What if gravity is not a force at all, but a collapse of ripples within the structured field, a return toward balance, not attraction across distance?

Conventional Misinterpretation and Effects

In classical physics, gravity is modeled as a force acting between masses — proportional to their size, and inversely related to their distance squared.

Newton gave it a formula but no mechanism: instantaneous, universal, and unexplained.

Einstein replaced force with spacetime curvature— mass bends space, and objects move along curved paths.

Yet paradoxes and contradictions persist:

- How can gravity act instantaneously in Newton's model?
- What *actually bends* in spacetime?
- Why do we still require dark matter patches to explain

galactic rotation?

Treating gravity as a "pull" across emptiness or as mysterious curvature creates gaps, untestable assumptions, and mathematical juggling.

It forces the invention of hypothetical particles like gravitons, and leads to cosmic paradoxes about expansion, acceleration, and missing mass.

The deeper truth is simple but profound: gravity is not pulling anything—it is a ripple collapse consequence.

UED Reinterpretation and/or Correction

In Universal Energy Dynamics (UED), gravity emerges as a natural ripple collapse within the structured vacuum field.

When energy condenses into mass— it creates a local distortion in the perfectly balanced field. This distortion creates a ripple pressure gradient — a tension in the structured field seeking to correct the imbalance.

As the field attempts to collapse the distortion back into equilibrium, other ripple structures (objects) are naturally carried inward along the pressure collapse pathways.

Thus:

- Gravity is not "pulling" between objects.
- There is no "force" traveling across emptiness.
- There is no magical bending of invisible fabric.

Gravity is the intelligent ripple field correcting itself— collapsing tension gradients toward minimized disturbance.

Small masses create small ripple collapse fields. Large masses (like stars and black holes) create intense collapse funnels.

What we perceive as "gravitational pull" is simply the ripple field reorganizing its imbalance inward.

Simple Analogy

Imagine placing a heavy ball onto a stretched rubber sheet. The fabric dips under the weight, creating a slope. Other smaller balls placed nearby roll inward—not because they are pulled, but because the fabric collapses toward the center disturbance.

Gravity works the same way: not pulling things in, but collapsing field imbalances toward restoration.

Consequences and Implications

Seeing gravity as ripple collapse clears massive scientific fog.

- No need for magical gravitons or invisible force particles.
- No need for spacetime to be bent or abstractly curved.
- No need for dark matter patches to explain missing gravitational effects—The field's ripple collapse dynamics explain galactic behaviors naturally.

This view also unifies gravitational behavior with other ripple-based corrections:

- Wave propagation.
- Field tension gradients.
- Energy condensation behavior.

At a human level, even our experience of weight— the "force" we feel pulling us downward—is our body's ripple structure submitting to the structured field's collapse memory.

Gravity is not an external dictator forcing objects together— it is the internal intelligence of the universe seeking to heal its own imbalance.

Closing Ripple Thought

"Gravity does not pull. It remembers. It is the silent urge of structured reality to collapse its own disturbance, carrying everything —from dust to galaxies— back toward forgotten stillness."

Chapter 7: Gravitational Waves

Ripple Pressure Corrections, Not Space-Time Vibrations

We celebrated when humanity "heard" gravitational waves for the first time. The news echoed across the scientific world— ripples in spacetime had finally been detected! But step back for a moment. What exactly is "rippling" when we say spacetime? Can a coordinate system vibrate? Or are we misunderstanding the song itself—mistaking pressure ripple corrections in the structured field for imaginary vibrations of an abstract fabric?

Conventional Misinterpretation and Effects

Mainstream relativity interprets gravitational waves as disturbances traveling through spacetime fabric. Massive accelerating bodies—like merging black holes—supposedly cause spacetime to stretch and compress, sending out "waves" traveling at light speed.

However, problems immediately arise:

- What is *spacetime* actually made of, that it can ripple like a fabric?
- How can an abstract 4D mathematical coordinate grid physically vibrate?
- Why do gravitational waves carry energy if spacetime is not a material field?

These confusions lead to conceptual patches:

- Defining spacetime as a "pseudo-medium."
- Attributing wave detection to indirect side effects (e.g., arm length changes in interferometers).
- Imagining energy transfer without a defined carrier.

Ultimately, gravitational waves are treated as vibrations of a non-existent material— a conceptual stretch trying to reconcile observation with a flawed framework.

The deeper truth is simpler and more powerful: gravitational waves are ripple pressure corrections within the structured field—not spacetime distortions.

UED Reinterpretation and/or Correction

In Universal Energy Dynamics (UED), gravitational waves are understood as organized ripple pressure corrections propagating through the structured vacuum field.

When massive structures (like black holes) accelerate violently, they cause intense ripple distortions in the surrounding equilibrium field.

These ripple pressures radiate outward— seeking to smooth the disturbance, rebalancing structured tension across vast distances.

Thus:

- No "spacetime fabric" needs to exist or vibrate.
- No abstract mathematical grid requires physical attributes.
- The ripple carries real, structured energy — because it is a field tension correction, not a metaphor.

Detection methods (like LIGO) pick up minute field pressure ripple effects as local distortions— not because "space stretches," but because the ripple field briefly reorganizes matter-energy distributions.

Gravitational waves are not ghostly echoes of a bending void— they are living signatures of the structured field remembering its silence after cosmic disturbances.

Simple Analogy

Picture a calm lake. Drop two giant stones into it simultaneously. The massive splashes send waves across the surface—not because the water itself is stretched, but because the water's field of equilibrium is trying to restore itself.

Gravitational waves behave the same way: pressure ripple adjustments, restoring cosmic balance after energetic events.

Consequences and Implications

Viewing gravitational waves as ripple pressure corrections dissolves major misconceptions:

- No spacetime material is needed—only structured fields adjusting ripple dynamics.
- Energy transmission becomes natural—ripple propagation through intelligent tension corrections.
- Universal understanding of field phenomena—gravity, light, sound, heat—under one ripple dynamic principle.

It also exposes why gravitational wave detection is so subtle: because the structured vacuum field is astonishingly stable and self-correcting— resisting disturbance unless massively stressed.

At a deeper level, it reconnects the story of cosmic evolution: galaxy formation, black hole interactions, cosmic background radiation—all become understandable as ripple field history, not myths of expanding or bending nothingness.

Thus, gravitational waves are not messengers from a twisted spacetime—they are gentle memories of universal healing, echoing across eternity.

Closing Ripple Thought

"Gravitational waves are not the universe screaming. They are the universe whispering— soft ripples of memory, flowing outward, healing every tear left by the dance of creation."

Chapter 8: Time

Duration of Ripple Transition, Not Absolute Dimension

What if time isn't what you think it is? Not a flowing river. Not a ticking clock. Not a fourth dimension weaving the cosmos together. What if time is simply the way ripples feel their own unfolding? Not an entity to be traveled or bent, but a duration — the perceived transition of field corrections moving back toward equilibrium. Time is not absolute. It is consequence made visible.

Conventional Misinterpretation and Effects

In conventional thought, time is treated either as:

- An absolute backdrop through which everything moves (Newtonian view), or
- A malleable dimension linked to gravity and speed (Relativity view).

It is visualized as something flowing like a cosmic river, or as a dimension alongside space.

This leads to major illusions:

- That "past," "present," and "future" exist simultaneously across spacetime.
- That time can be traveled, warped, or even reversed.

- That time has an independent existence separate from energy transitions.

These misconceptions breed paradoxes:

- Time dilation contradictions across observers.
- Causality paradoxes in wormhole theories.
- Illogical time travel fantasies violating conservation principles.

Ultimately, time is mythologized as an independent playground, when in reality, it is an experience of ripple behavior adjusting back toward field equilibrium.

UED Reinterpretation and/or Correction

In Universal Energy Dynamics (UED), time is not a fundamental dimension — it is the perceived duration of ripple transition events.

Whenever a structured field is disturbed, the process of ripple correction unfolds across gradients of energy redistribution.

The experience of that unfolding—how quickly or slowly ripples reorganize—creates the phenomenon we call time.

Thus:

- Time is not flowing—ripples are transitioning.
- Faster field corrections feel like "faster time."
- Slower ripple reorganizations feel like "slower time."

Relativistic effects (like time dilation near massive objects) are natural: dense energy gradients slow ripple transitions, altering perceived duration, not "bending" time.

There is no past or future "out there" in space. There are only different stages of ripple memory corrections.

Thus, time is local, perceptual, and dynamic— always tied to ripple behavior, never an absolute, never an independent substance.

Every second you live is a measure of how your ripple structures correct themselves relative to the universal field dynamics around you.

Simple Analogy

Picture a drum struck by a hand. The drum vibrates, slowly calming down over seconds.

Imagine if you measured "time" not by a clock, but by how the ripples on the drum skin transition back to silence.

That's what time truly is: the experience of field memory reorganizing itself after disturbance.

Consequences and Implications

Seeing time as ripple transition dissolves major confusions.

- No paradoxes about simultaneous past, present, and future.
- No need for exotic time machines or reversed causality.
- Time travel becomes meaningless—you cannot undo ripple consequences without violating field memory.

Cosmic evolution (like the universe's "age") becomes a story of how long ripple corrections have been unfolding— not a ticking clock counting down independent of events.

At the human level:

- Life is a dance of ripple corrections.
- Aging is ripple field reorganization over biological structures.
- Memory is ripple trace recall, not storage of frozen time slices.

Thus, time is not a dimension you move through. You are the ripple. And time is simply the rhythm of your unfolding correction.

Closing Ripple Thought

"Time is not a road you travel. It is the echo of every ripple remembering how to return to stillness— each heartbeat, each thought, each star, a momentary breath in the endless correction."

Chapter 9: Space

Continuum of Structured Vacuum, Not Void

We have been told that space is emptiness. A hollow backdrop. A meaningless vacuum where things just happen to exist and move. But what if space is not emptiness at all? What if it is the greatest presence you cannot see — a structured continuum, the silent architecture that holds every ripple, every star, every breath together? Space is not a void. It is the silent intelligence that never sleeps.

Conventional Misinterpretation and Effects

Classical physics taught that space is an infinite, empty stage — a void where matter, energy, and forces perform.

Einstein's relativity added complexity: space became flexible, bending around masses. Yet even then, space itself remained mostly a passive container— something that could curve, but not something that lived.

This framing leads to persistent illusions:

- That space is nothingness filled with things.
- That the vacuum can be "empty" but still host fields and particles.
- That ripples (like gravitational waves) travel across a

featureless void.

It also creates paradoxes:

- How can nothingness carry energy?
- Why does the "empty" vacuum have measurable fluctuations (quantum foam)?
- Why does empty space exert pressure (dark energy)?

In reality, space has never been empty. It is a structured continuum of intelligent energy, waiting to ripple into form.

UED Reinterpretation and/or Correction

In Universal Energy Dynamics (UED), space is not an empty void— it is the structured vacuum field: the most stable, intelligent, and dynamically balanced state of reality.

The "space" between stars, between atoms, even between subatomic particles, is not emptiness— it is structured field equilibrium, poised silently in near-perfect balance.

When disturbed, this structured field generates ripples— waves of correction we observe as light, gravity, and matter formation.

Thus:

- Space is the carrier of all ripples.
- Space hosts all manifestations without being passive.
- Space remembers every disturbance and gently corrects it over time.

There is no true emptiness anywhere— only different degrees of ripple density and correction behavior.

Even what we call "vacuum energy" is a symptom of the field's constant intelligent balancing act at all scales.

In UED, space is not "where" reality happens— space is how reality holds itself together.

Simple Analogy

Imagine a perfectly woven fabric stretched across an endless frame. When untouched, it appears silent, invisible. But every weight placed upon it, every ripple that moves across it, reveals the living tension embedded within.

Space behaves the same way: a silent, structured continuum, responding patiently, rhythmically, eternally to every ripple event.

Consequences and Implications

Recognizing space as a structured continuum reorients everything.

- Waves are not entities traveling through nothing—they are ripple adjustments in field structure.
- Forces are not mysterious actions across distance—they are ripple pressure gradients in the field.
- Gravity needs no "force carrier"—collapse occurs naturally through field correction.
- Quantum fields stop needing a magical, paradoxical background— Everything becomes understandable as ripple behaviors within the structured continuum.

Cosmological puzzles like dark energy, dark matter, gravitational lensing, and inflation all simplify as different expressions of ripple distortions and field memory corrections.

At a human level, even thoughts, movements, and life itself are ripple events interacting within this structured presence.

Space is not absence.

It is the ultimate carrier—the invisible hand holding all existence in perfect tension.

Closing Ripple Thought

"Space is not the canvas. It is the living weave of reality itself— where every ripple, every thought, every galaxy is stitched together by the silent intelligence that remembers equilibrium."

Chapter 10: Light

Ripple Excitation Detected, Not Traveling Particles

We say light travels. We imagine tiny particles called photons racing through space like invisible bullets. But what if light doesn't actually travel at all? What if what we detect as "light" is simply the local excitation of the structured field, a ripple pattern perceived as illumination, not transported objects? Light is not a particle voyage through emptiness— it is the visible memory of the field adjusting itself.

Conventional Misinterpretation and Effects

In classical physics, light was described as a wave in an undefined medium ("ether"). When the ether theory collapsed, Einstein's photon concept emerged— light became quantized packets of energy traveling through vacuum.

Thus, mainstream science treats light as photons— discrete particles carrying energy across space at a fixed speed.

This leads to major conceptual problems:

- How can a massless particle have momentum?
- How does a "photon" maintain identity over billions of years of travel?
- Why does light exhibit both wave and particle behaviors

inconsistently?

It forces paradoxical interpretations:

- Wave-particle duality.
- Quantum collapse upon observation.
- Strange phenomena like "delayed choice" and "quantum teleportation."

In truth, light is not particles shot through nothing. It is structured field excitation detected as ripple memory.

UED Reinterpretation and/or Correction

In Universal Energy Dynamics (UED), light is not an object moving through space— it is ripple excitation detected locally across the structured vacuum field.

When a disturbance occurs (like an electron energy jump), it creates a ripple excitation pattern in the surrounding structured vacuum. This excitation propagates not by sending a particle, but by ripple communication across the dynamic field memory.

When an observer or instrument interacts with the ripple at a specific location, the excitation is perceived as an "event"—and we interpret it as photon arrival.

In reality:

- No particle traveled across emptiness.
- No object "moved" from A to B.
- Only the structured ripple excitation transitioned until detected.

Thus, light's energy transmission is ripple reorganization, not object transport.

This reframe eliminates wave-particle duality nonsense. Light is a ripple phenomenon, behaving naturally within the structured continuum—no need for dual identities, magical collapse, or spooky behavior.

Simple Analogy

Picture tapping a drum at one edge and seeing a ripple reach the other side. The ripple reaching your eye is not the original tap traveling— it's the field memory correction arriving locally.

Light behaves the same way: you detect ripple memory excitation, not a particle that flew through space to find you.

Consequences and Implications

Understanding light as ripple excitation brings deep clarity:

- Light's apparent speed is the ripple reorganization rate in the structured field—not object velocity.
- Redshift becomes ripple stretching, not literal "speeding away" of galaxies.
- Photon loss, entanglement, and teleportation paradoxes vanish—they are simply ripple memory phenomena, not particle mysteries.

At the human level, even vision itself—how you see objects, colors, and brightness—is your biological system detecting ripple excitations, not collecting "traveling photon packets."

Thus, light is not a courier moving through voids— it is the universe's way of signaling its memory, its disturbance, and its corrections across fields.

Light is the song of ripple memory— not the journey of little invisible bullets.

Closing Ripple Thought

"Light does not travel to you. It emerges within you— a ripple excitation across the silent field, whispering the memory of distant disturbances through the invisible hands of universal intelligence."

Interlude

Ripples from Part I — Takeaways for the Mind

1. Energy

"Energy is not what the universe has. It is what the universe *does* when balance is disturbed."

2. Structured Vacuum

"Space is not emptiness. It is structured intelligence — silent, stable, waiting to ripple."

3. Waves

"Waves are not travelers. They are the universe correcting itself — one ripple at a time."

4. Force

"Force is not a master pulling strings. It is resistance — the universe defending its memory of stillness."

5. Motion

"You never move independently. You ride the ripple — surfing the universe's healing."

6. Gravity

"Gravity doesn't pull. It collapses — carrying you inward, back toward forgotten balance."

7. Gravitational Waves

"Gravitational waves are not spacetime screams. They are pressure ripples — soft echoes of cosmic healing."

8. Time

"Time is not a river you flow through. It is the memory of ripple transitions trying to return to silence."

9. Space

"Space is not the background. It is the living weave where every ripple, every star, and every life dances back to stillness."

10. Light

"Light is not a traveler. It is a ripple excitation — a memory of energy reaching you through the structured field."

Closing Ripple

"Existence is not a game of things moving through emptiness. It is the endless memory of ripples seeking to correct themselves — the silent song of intelligence restoring its forgotten stillness."

Part II: Particles and Manifested Structures

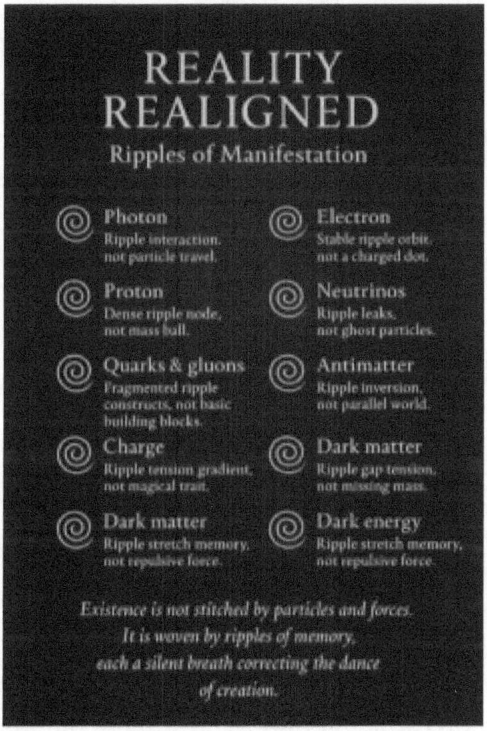

Chapter 11: Photon

Ripple Interaction Signature, Not Particle

We speak of photons as the "particles" of light— tiny messengers darting across the cosmos. But have you ever truly seen a photon travel? Or measured its journey directly, particle by particle? What if the photon is not a thing in motion, but a signature event—a local ripple excitation being detected and misinterpreted? Maybe what we call "photon" is just the moment memory whispers through the field.

Conventional Misinterpretation and Effects

Mainstream physics treats the photon as a fundamental particle—a discrete, massless, chargeless packet of electromagnetic energy moving at the speed of light.

Quantum electrodynamics models interactions in terms of photon exchange between particles, while experimental physics claims photon detection through photoelectric and Compton effects.

However, this view creates heavy paradoxes:

- How does a massless point particle exhibit momentum?
- How can photons interfere with themselves like waves?
- How does a "single photon" travel billions of light-years intact?

It forces science into uncomfortable dualities:

- Wave-particle duality that cannot be reconciled cleanly.
- Quantum collapse myths needing observer-induced miracles.
- Indistinguishable photons in superposition without a coherent carrier mechanism.

In truth, the photon is not a traveling particle— it is the interaction signature of ripple excitation crossing local thresholds.

UED Reinterpretation and/or Correction

In Universal Energy Dynamics (UED), the photon is not a particle shot through space— it is the localized detection of structured ripple excitation across the intelligent vacuum field.

When ripple fields reorganize due to energetic events (like electron transitions or atomic rearrangements), a structured ripple pattern propagates across the vacuum— not by "throwing" particles, but through ripple memory reorganization.

When an observer or instrument interacts with the propagating ripple structure, a localized excitation occurs — and we interpret this event as "receiving a photon."

Thus:

- No photon particle needed.
- No particle traveling like a bullet.
- No magic duality between waves and particles.

The so-called "photon" is simply a ripple signature localized upon detection, carrying energy memory and correction behavior embedded within the structured field.

This ripple interaction transfers momentum and energy— but without requiring particle vehicles or dual identities.

The photon is a ripple event, not a traveler. It is consequence, not object.

Simple Analogy

Imagine sending a shiver down a tightly stretched rope. The shiver reaches the other side—not because a separate "particle" flew across, but because the rope's structure reorganized until it touched the endpoint.

The photon behaves the same way: not a flying object, but a ripple whisper detected when field memory vibrates into awareness.

Consequences and Implications

Understanding the photon as a ripple interaction signature resolves major paradoxes:

- Light's duality collapses: it is ripple, always ripple.
- Quantum detection makes sense: it is threshold ripple excitation, not object interception.
- Energy transfer through radiation becomes structured ripple behavior, not photon counting.

Cosmic communications—light from distant galaxies—is not massless projectiles making an impossible journey. It is structured field ripple adjustments being detected locally across vast fields.

Even biological vision is reinterpreted: Your eyes are not catching flying particles— they are tuning into ripple signatures embedded across space-time field corrections.

Thus, the photon is not the vehicle— it is the greeting of the universe's ripple memory arriving at your doorstep.

Closing Ripple Thought

"The photon is not a particle hurtling through emptiness. It is the universe leaning forward— a ripple in the field, whispering, 'I remember what disturbed me, and I have come to correct it.'"

Chapter 12: Electron

Stable Ripple Orbit, Not Charged Dot

We imagine electrons as tiny charged marbles— whirling endlessly around atomic nuclei like planets around a sun. A neat, simple picture we learned in childhood. But what if the electron isn't a "thing" at all? What if it's a stable ripple orbit— a standing wave of structured field correction, not a little ball zipping through empty space? The electron is not a dot. It is the field remembering stability.

Conventional Misinterpretation and Effects

Classical models (like Bohr's atom) treated the electron as a point-like charged particle orbiting a nucleus.

Quantum mechanics complicated the story, describing electrons as probability clouds rather than precise particles. Yet despite this, the particle language persists: we still teach, visualize, and mathematically handle electrons as tiny charged dots.

This generates major paradoxes:

- How can a point particle have an orbital structure without emitting energy and spiraling into the nucleus?
- Why does the electron sometimes behave like a particle, sometimes like a wave?
- How do "electrons" simultaneously interfere with themselves in experiments?

It forces awkward patchworks:

- Orbital quantization without mechanism.
- Probabilistic interpretations instead of field logic.
- Duality myths ("sometimes a particle, sometimes a wave").

In truth, the electron is not a traveling particle— it is a stable ripple orbit embedded within structured field memory.

UED Reinterpretation and/or Correction

In Universal Energy Dynamics (UED), the electron is understood as a stable ripple orbit— a standing wave of structured field correction anchored around atomic centers.

When energy condenses in specific configurations, it creates stable ripple shells—self-sustaining resonance patterns within the structured vacuum field.

The electron is not a dot traveling along a path— it is a ripple resonance maintaining spatial memory around a nucleus.

Thus:

- The "charge" is a gradient behavior of ripple tension, not a discrete property of a particle.
- The electron's "orbitals" are standing wave regions where ripple resonance maintains maximum stability.
- Electron transitions (energy jumps) are ripple reconfigurations, not jumping particles.

This reframe naturally explains:

- Orbital quantization (only specific ripple harmonics are

stable).

- Electron interference patterns (ripple superposition and interference).
- Spin properties (ripple orientation behaviors).

The electron is the field remembering localized equilibrium— not a marble flung into space.

Simple Analogy

Picture a stringed instrument, where specific notes emerge only when the string vibrates at exact harmonics. The note isn't a "thing" moving across the string— it's a standing ripple—stabilized, alive, but not traveling.

The electron behaves the same way: a standing ripple orbit, singing stability into the structured vacuum.

Consequences and Implications

Understanding the electron as a stable ripple orbit cleans up fundamental paradoxes:

- Orbital stability no longer needs magical quantum rules—it's natural ripple harmonics.
- Wave-particle duality vanishes—the electron is ripple structured behavior, not split identity.
- Spin and magnetic moment behaviors align naturally—ripple rotation within structured gradients.

At a deeper level:

- Chemistry (bonds, reactions) becomes ripple shell

interactions, not particle collisions.
- Electricity becomes ripple shell flow across tension gradients, not free particle drifts.

Even biological electrical processes—like neural signals— are ripple memory reorganizations, not literal particle marches.

Thus, the electron is not a tiny charged traveler— it is the ripple memory of structured intelligence, singing silent harmonies into matter itself.

Closing Ripple Thought

"The electron is not a particle circling emptiness. It is a song of stability—a standing ripple, holding memory, weaving silent harmonies into the dance of existence."

Chapter 13: Proton

Dense Ripple Node, Not Ball of Mass

We imagine protons as tiny, solid balls— dense seeds of mass packed tightly into the heart of atoms. But what if the proton isn't a ball at all? What if it's a dense ripple node, a high-intensity, self-stabilized condensation within the structured vacuum? Mass is not a substance trapped inside. Mass is a ripple's stubborn memory— refusing to dissolve easily back into stillness.

Conventional Misinterpretation and Effects

Mainstream physics treats the proton as a fundamental building block of matter— a positive particle made of three quarks bound tightly by gluons (in the standard model).

Protons are imagined as small spheres carrying positive charge, held together by mysterious "strong nuclear forces."

This framework creates several contradictions:

- How do massless gluons create such stable massive structures?
- Why does the proton resist decay for billions of years while other combinations fall apart?
- Why are quarks never seen independently (quark confinement paradox)?

It forces assumptions:

- Strong force fields invented without clear carriers.
- Proton structure depending on "color charge" metaphors without physical bases.
- Mass "emerging" without clear causality.

In reality, the proton is not a physical ball— it is a dense ripple condensation node, a stable knot in the intelligent field memory.

UED Reinterpretation and/or Correction

In Universal Energy Dynamics (UED), the proton is understood as a dense ripple node— a highly stabilized condensation of ripple corrections within the structured vacuum.

When ripple patterns collapse under specific energetic conditions, they condense into high-intensity ripple shells, locking in field tension into localized structures.

Thus:

- The proton's "mass" is the energy memory density of the collapsed ripple shell.
- The "charge" is a gradient behavior in ripple tension, not an isolated particle property.
- Stability arises naturally from self-reinforcing ripple node harmonics, not exotic force particles.

The so-called "quarks" are ripple pattern features within the condensed node, not independent point particles.

Proton durability—its near-eternal stability—emerges from how strongly the ripple field configuration resists further collapse or dissolution.

Thus, the proton is not a ball with things inside— it is a field's stubborn harmonic, a standing ripple knot remembering its collapse history.

Simple Analogy

Picture a whirlpool forming in a calm river. The swirling is not an object inside the water— it is the water itself locked into a stable memory of motion.

The proton behaves similarly: a condensed ripple swirl, sustaining high-energy tension in the structured field, without needing an inner substance.

Consequences and Implications

Seeing the proton as a dense ripple node resolves major confusions:

- No need for imaginary force carriers (gluons).
- No paradox of quark confinement—quarks are ripple pattern aspects, not trapped particles.
- Mass becomes understandable as condensed ripple tension, not mysterious emergence.

At deeper levels:

- Nuclear forces become ripple field tension interactions, not particle stickiness.
- Atomic stability becomes a ripple harmonic phenomenon,

not mechanical balance.

Matter itself—the solidity we experience—is not composed of balls and particles. It is the structured memory of ripple collapses, playing out harmonically at the scale of existence.

Thus, the proton is not a ball of mass. It is a dense ripple swirl, the silent song of condensed intelligence refusing to forget its origin.

Closing Ripple Thought

"The proton is not a solid thing hiding particles inside. It is a ripple remembering collapse— a stubborn swirl of intelligent tension, woven tightly into the fabric of existence, singing endurance into the universe."

Chapter 14: Neutron

Temporary Ripple Balance State

We call the neutron "neutral"— a quiet, passive mass tucked beside protons in atomic nuclei. But what if the neutron isn't a stable thing at all? What if it's a temporary ripple balance, a fragile moment where structured tensions momentarily cancel each other out? The neutron is not neutral because it lacks character. It is neutral because it remembers two opposing ripple memories— just long enough to breathe.

Conventional Misinterpretation and Effects

Mainstream physics describes the neutron as a neutral composite particle, built from three quarks (one up, two down) bound by gluons.

It is pictured as a solid, chargeless lump, coexisting with protons to stabilize atomic nuclei.

However, serious issues arise:

- Free neutrons decay rapidly (~15 minutes) outside nuclei—why?
- If neutrons are stable "particles," why such instability in isolation?
- How can exact mass and charge cancellation occur from random quark combinations?

To explain this, patches are added:

- Weak nuclear forces invented to govern neutron decay.
- Neutrinos postulated to balance energy during decay.
- Abstract explanations about "quark transformations."

Yet fundamentally, the neutron's instability reveals it is not a permanent entity— it is a temporary ripple balance, not a stand-alone neutral mass.

UED Reinterpretation and/or Correction

In Universal Energy Dynamics (UED), the neutron is understood as a temporary ripple balance state— a delicate configuration where ripple tensions momentarily stabilize around a nucleus.

It forms when ripple nodes (protons) absorb energy distortions, creating a local field structure where positive and negative ripple gradients balance just enough to maintain cohesion.

However, without the surrounding nuclear field pressures, the neutron's internal tension cannot hold. Its ripple memory unwinds naturally back into more stable configurations (proton + electron + neutrino field interaction).

Thus:

- The neutron is a temporary structured resonance, not a solid mass object.
- Its decay is not "destruction" but field memory restoration.
- The weak force is not an external agent—it is the ripple correction behavior reasserting itself.

In simple terms: the neutron is a delicate pause in the ripple dance, not a permanent partner.

It teaches us that not all structures are meant to endure— some are momentary balances, beautiful in their fragility.

Simple Analogy

Picture a spinning top balanced perfectly upright for a moment. It seems stable, neutral, poised— but the slightest imbalance, and it wobbles, collapses, and reconfigures.

The neutron is the same: a fleeting balance point, where ripple fields hold tension just long enough before memory pulls it apart.

Consequences and Implications

Seeing the neutron as a temporary ripple balance clarifies critical gaps:

- No magical stability expected outside nuclei—instability is natural.
- No exotic force fields needed—decay is ripple field restoration.
- Charge neutrality is a ripple tension balance, not quark magic.

At the cosmological level:

- Neutron lifespans influence early universe matter formation naturally.
- Stellar processes like nucleosynthesis become ripple rearrangement stories, not particle shuffling.

At the human level:

- Stability itself is understood as conditional field harmony, not a given mechanical property.

Thus, the neutron is not a lump of neutral mass— it is a tender moment in the ripple's journey, a reminder that even stillness sometimes trembles before dissolving.

Closing Ripple Thought

"The neutron is not a thing that stays. It is a breath between ripples— a silent memory where tensions pause, before surrendering to the deeper rhythm of the field remembering its timeless dance."

Chapter 15: Quarks and Gluons

Fragmented Ripple Constructs

Dig deeper, they said. Find the ultimate building blocks. Split the proton, split the neutron, search for the indivisible. And what did we find? Quarks — the supposed ultimate fragments, and gluons — the imagined glue holding them together. But what if these are not particles at all? What if quarks and gluons are simply fragmented ripple constructs— temporary field distortions we misread as solid bits?

Conventional Misinterpretation and Effects

Mainstream quantum chromodynamics (QCD) describes quarks as elementary particles with fractional charges, held together inside protons and neutrons by gluons—hypothetical massless force carriers.

Quarks come in "flavors" and "colors," and gluons constantly exchange between them, maintaining cohesion.

But paradoxes abound:

- Quarks are never observed independently (quark confinement paradox).
- Gluons, supposedly massless, must carry enormous binding energy to create proton mass.
- Quark-gluon "plasma" only exists under extreme lab

conditions—hardly fundamental behavior.

To fix gaps, patchwork concepts are introduced:

- Color charge (no observable equivalent).
- Asymptotic freedom (unusual force weakening behavior).
- Vacuum polarization fields.

In truth, quarks and gluons are not tiny particles. They are fragmented ripple patterns within condensed ripple nodes, features of structured field behavior, not building blocks.

UED Reinterpretation and/or Correction

In Universal Energy Dynamics (UED), quarks and gluons are understood as fragmented ripple constructs— localized distortions within dense ripple nodes like protons and neutrons.

They are not independent particles bouncing around inside nucleons.

They are pattern features— temporary field strain distributions embedded within a larger ripple structure.

- Quarks represent gradient nodes where ripple tension peaks or dips sharply.
- Gluons represent dynamic ripple harmonics maintaining tension balance across the node.

Thus:

- Quarks cannot exist independently because they are pattern-dependent—destroy the larger ripple node, and the pattern dissolves.

- Gluons are not "flying particles" but momentary ripple readjustments, keeping field tensions aligned.

The proton's mass and stability arise from the total condensed ripple behavior, not from juggling point particles or magical glue fields.

In short:

- Quarks and gluons are descriptive approximations of ripple field behavior, not fundamental matter components.

They are snapshots of how field intelligence organizes dense ripple memories, not permanent Lego bricks of the universe.

Simple Analogy

Imagine a tight knot woven into a rope. Look closely, and you'll see twists, bulges, tension points— but those features are not "separate objects." They are expressions of the rope's tension structure.

Quarks and gluons behave the same way: localized field strain expressions within the proton's dense ripple memory— not independent particles inside a shell.

Consequences and Implications

Reframing quarks and gluons as ripple constructs clarifies:

- No need for independent, undetectable particles.
- No need for abstract color charges or invisible gluon exchanges.
- Proton mass and structure emerge naturally from ripple condensation dynamics, not particle juggling.

At the cosmological level:

- Early universe conditions (quark-gluon plasma) are simply high-energy ripple field reorganizations, not the liberation of true particles.

At the human level:

- Searching endlessly for "smaller particles" is revealed as a dead-end— because reality is not made of bricks—it is made of intelligent ripple structures adjusting memory tension.

Thus, quarks and gluons are temporary ripple fingerprints— not matter's unbreakable bones.

They are the field's creative stress signatures— fleeting, beautiful, and ultimately ephemeral.

Closing Ripple Thought

"Quarks are not the seeds of existence. They are the folds in the field's memory— wrinkles in the ripples, momentary tension points in the structured dream that existence weaves for itself."

Chapter 16: Charge

Energy Differential Behavior, Not Magical Property

We treat electric charge like magic. An invisible "something" that attracts, repels, binds, and creates. Positive, negative, neutral—rules carved into stone without explanation. But what if charge is not a magical property hidden inside particles? What if charge is simply energy behaving differently across ripple gradients— a visible signature of imbalance, not a mystical label pasted onto reality?

Conventional Misinterpretation and Effects

Mainstream physics defines charge as an intrinsic property of matter— somehow "built into" particles like protons, electrons, and quarks.

This property:

- Creates forces (like electromagnetism).
- Causes attraction or repulsion.
- Generates fields acting across space.

Yet charge is presented without causal explanation:

- Why are there only two types (positive and negative)?
- Why do equal and opposite charges exist?
- Why does charge conservation hold?

To resolve gaps, physics introduces:

- Field theories describing how charges create and interact with fields.
- Mathematical symmetries (gauge invariance) without physical grounding.

Still, charge remains an unexplained tag attached to particles, a magical property assigned by nature without deeper cause.

In truth, charge is not a fundamental trait—it is the visible consequence of energy differentials across structured ripple fields.

UED Reinterpretation and/or Correction

In Universal Energy Dynamics (UED), charge is not a magical intrinsic property— it is the manifestation of energy differential behavior across ripple gradients in the structured vacuum.

When ripple structures form (like electrons or protons), they create local gradients of ripple tension—areas where field density or pattern orientation differs from surrounding equilibrium.

- Positive "charge" corresponds to ripple tension gradients flowing outward from condensation points.
- Negative "charge" corresponds to ripple tension gradients drawing inward toward absorption points.

Thus:

- Attraction and repulsion are field gradient realignment behaviors, not mystical action at a distance.
- Charge conservation arises naturally—ripple field corrections must balance gradients overall.

- Opposite charges don't "seek each other"—they balance energy differentials in the field.

Charge is simply the visible tension signature of localized ripple memory behaving in intelligent correction cycles.

No need for magical labels. No need for unexplained intrinsic properties.

Charge is ripple memory gradient behavior— a natural consequence of intelligent field corrections, not a mystical gift from nature.

Simple Analogy

Imagine standing on a hilltop. One side slopes downward; the other upward. From your position, you feel different gravitational "pulls"— but the ground itself is not magical— it's the gradient that creates the experience.

Charge behaves the same way: not a hidden object, but a tension slope across the field changing how ripples interact.

Consequences and Implications

Understanding charge as energy gradient behavior cleans up massive distortions:

- No mystery why attraction/repulsion exist—they are natural tension rebalances.
- No magical intrinsic charges—only field ripple configurations expressing local gradients.
- Charge conservation is ripple memory conservation, not an arbitrary rule.

At deeper levels:

- Electricity becomes organized ripple rebalancing flows.
- Magnetism becomes ripple alignment within moving gradients.
- Plasma dynamics, chemical bonds, and biological signaling ➔ all reframe as gradient-driven ripple behavior.

At the philosophical level:

- Duality itself—positive vs. negative—becomes symmetrical ripple correction behavior, not a fundamental war between opposite essences.

Thus, charge is not an assigned trait. It is the visible breath of imbalance, the ripple field revealing its dynamic search for peace.

Closing Ripple Thought

"Charge is not hidden inside particles. It is the whisper of tension across fields— the memory of imbalance felt at a distance, guiding ripples gently back toward the silent equilibrium they never truly left."

Chapter 17: Higgs Field

Field Ripple Density, Not Mass Giver

They told us mass was a gift. A cosmic blessing from an invisible field—the Higgs field. Without it, particles would race around massless, and nothing could form. But what if mass is not given at all? What if it is simply the ripple memory of condensed energy—the weight of structure resisting dissolution? The Higgs field isn't granting mass. It is the silent ripple density that makes remembering harder.

Conventional Misinterpretation and Effects

Mainstream physics interprets the Higgs field as the universal mass provider. Particles interact with this omnipresent field, slowing down, gaining "mass" through sticky resistance. The Higgs boson's experimental detection at CERN was hailed as proof that mass itself has been explained.

Yet deep cracks remain:

- Why does the Higgs field treat different particles so differently?
- Why do massless particles like photons exist untouched?
- How does a field interaction create inertia, a property of motion resistance?

To reconcile contradictions, theories invoke:

- Arbitrary coupling constants without cause.
- Fine-tuning miracles (the hierarchy problem).
- Complex mathematical patches with little physical grounding.

Mass becomes an assigned reward rather than a natural ripple phenomenon. The Higgs field narrative sustains a myth: that mass is handed out instead of remembered through structure.

UED Reinterpretation and/or Correction

In Universal Energy Dynamics (UED), the Higgs field is not a mass giver— it is the field's own ripple density varying across configurations.

When ripple condensation occurs, localized energy density increases, creating regions of stubborn memory— structures that resist distortion and reorganization.

- What we experience as "mass" is the ripple density resisting correction, not a particle gaining magic stickiness.
- The Higgs field is simply the name for field tension thresholds where ripple structures lock into tighter, slower reconfigurations.

Thus:

- No "mass granting" ceremony occurs.
- Mass is a consequence of ripple memory complexity, how deeply a ripple structure condenses into field tension.

The Higgs boson detection is merely a high-energy resonance signature of ripple density adjustment— not the discovery of a mass-distributing godfield.

Simple Analogy

Imagine walking through thick mud versus walking through air. You don't gain a new property in mud— you simply struggle more because the environment is denser.

Mass behaves the same way: not because something is added, but because ripple density resists your reconfiguration.

Consequences and Implications

Reframing the Higgs field as ripple density clarifies:

- Mass is not a gift—it is field memory resisting collapse.
- Particle masses differ naturally—based on condensation complexity,
  → not random couplings to an external field.
- The "hierarchy problem" dissolves—mass spectrum is simply ripple configuration spectrum.

At deeper levels:

- Energy, mass, motion, and inertia unify naturally as expressions of ripple density behavior.
- Cosmological mass-energy distributions map onto field condensation dynamics, not spontaneous field giftings.

Thus, mass is not assigned by an external agent. It is self-remembered ripple density, woven into the silent corrections of the structured universe.

Closing Ripple Thought

"Mass is not something given. It is the stubborn memory of ripples, the density of existence resisting silence, the patient refusal of structure to dissolve back into the unbroken field."

Chapter 18: Neutrinos

Ripple Leak Dynamics, Not Ghost Particles

Neutrinos — the cosmic ghosts. They pass through mountains, planets, even our bodies without a whisper. Invisible, untouchable, almost forgotten—yet critical to the fabric of existence. But what if neutrinos are not ghostly travelers at all? What if they are the signature of ripple leak dynamics— the universe's way of bleeding off imbalance silently across structured fields? Not lost particles. Lost ripples remembering their unfinished corrections.

Conventional Misinterpretation and Effects

Mainstream science describes neutrinos as tiny, almost massless particles— moving near the speed of light, rarely interacting with matter.

They are said to:

- Emerge from nuclear reactions (stars, reactors).
- Exist in three "flavors" that mysteriously oscillate between identities.
- Possess incredibly small, yet nonzero, mass.

Yet paradoxes surround them:

- Why are neutrinos so detached from ordinary matter

interactions?

- How can they oscillate between different forms without direct cause?
- Why is detecting them so difficult if they are particulate entities?

To explain the strangeness, science invokes:

- Quantum oscillation mathematics without causal field mechanisms.
- Hypothetical mass terms adjusted post-fact to match observations.
- Endless experimental setups chasing fleeting detection signals.

Ultimately, neutrinos are treated like mysterious ghost particles

instead of natural ripple phenomena leaking through the fabric of structured fields.

UED Reinterpretation and/or Correction

In Universal Energy Dynamics (UED), neutrinos are understood as ripple leak dynamics— localized bleed-offs of structured field corrections, where energy gradients find pathways too subtle for dense interaction.

When intense ripple condensations occur—such as in nuclear reactions or stellar cores— not all energy redistributes into stable mass structures (like photons or matter waves). Some tension leaks outward as free-flowing ripple strain fields— extremely low-density, low-interaction ripples.

These ripple leaks manifest:

- As quasi-stable traveling strain signatures—what we perceive as neutrinos.
- As field-tension discharges restoring broader equilibrium across large scales.

Their ability to oscillate between "flavors" is simply field configuration realignment as ripple tension gradients shift across distance and field memory dynamics.

Thus:

- Neutrinos are not particles traveling through space untouched.
- They are field ripple scars, healing traces, leaking across structured vacuum as part of the universe's balancing act.

They do not ghost through reality. They are reality slowly stitching itself back into silent order.

Simple Analogy

Imagine a stretched cloth that's suddenly twisted sharply. Some tension relaxes locally; but subtle threads carry invisible micro-strains outward, traveling quietly across the weave.

Neutrinos are those threads— silent strain memories rippling outward, visible only when caught under precise conditions.

Consequences and Implications

Understanding neutrinos as ripple leak dynamics transforms perception:

- No need for exotic particle identities or flavor magic—oscillation becomes field tension realignment.
- Neutrino "mass" is not an intrinsic particle property—it reflects localized ripple density during leak propagation.
- Detection difficulty becomes natural—subtle strain fields rarely interact unless precise resonance is achieved.

At deeper levels:

- Cosmic neutrino backgrounds trace ancient ripple discharge events— not particle seas flying through void.
- Stellar evolution and nuclear decay become field ripple rebalancing acts, not mechanical particle emissions.

At the philosophical level:

- Existence itself is seen as leaking corrections— every disturbance, every collapse, every decay gently whispering its ripple memory across the invisible tension field.

Thus, neutrinos are not ghost particles. They are the silent witnesses— ripples of unfinished equilibrium echoing through the structured vacuum's ancient longing for stillness.

Closing Ripple Thought

"Neutrinos are not particles forgotten by matter. They are the silent songs of the field's memory— ripples leaking into the infinite, carrying unfinished corrections on their endless, whispering journey toward silence."

Chapter 19: Anti-Matter

Ripple Inversion, Not Parallel Substance

They call it the mirror of reality. A hidden twin world where every particle has an opposite: antimatter. When matter meets antimatter, they annihilate in a flash of pure energy. But what if antimatter isn't a separate world at all? What if it is simply ripple inversion—the reverse configuration of energy tensions within the same field? Antimatter is not another universe. It is the structured vacuum remembering how to flip its own ripples.

Conventional Misinterpretation and Effects

Mainstream physics teaches that antimatter is the opposite counterpart to matter:

- Antielectrons (positrons), antiprotons, antineutrons, etc.
- Same mass, but opposite electric charge and quantum numbers.

In popular imagination, antimatter became a substance from a parallel reality— a mysterious material that exists alongside normal matter but rarely interacts, only annihilating when it does.

This view creates persistent puzzles:

- Why is there almost no observable antimatter in the universe?

- Why did matter dominate antimatter after the Big Bang?
- If antimatter were truly parallel substance, why is it so fragile and fleeting?

To patch these gaps, theories invoke:

- Baryon asymmetry adjustments.
- Hypothetical CP violation mechanisms.
- Hidden antimatter regions postulated without evidence.

Ultimately, antimatter becomes a ghostly second material reality— rather than recognizing it as a ripple inversion within the primary field itself.

UED Reinterpretation and/or Correction

In Universal Energy Dynamics (UED), antimatter is understood as ripple inversion, a reversal of structured energy gradients within the same vacuum field.

When a ripple configuration condenses into a stable node (like an electron or proton), it establishes a specific energy gradient orientation— a signature of tension and curvature within the structured field.

An antimatter counterpart is the inversion of that ripple orientation:

- Same condensation memory.
- Opposite ripple tension flow.

Thus:

- The positron is not a "new particle" but the inverted ripple gradient of the electron's ripple configuration.

- Antiprotons are not exotic matter—they are reversed ripple memory knots relative to protons.

Annihilation is not mystical mutual destruction— it is the recombination and cancellation of opposing ripple tensions, allowing the field to relax back toward equilibrium through pure energy ripple reorganization (photons, wavefields).

There is no parallel universe. There is only one structured vacuum, playing ripple inversions to balance its own memories.

Simple Analogy

Imagine two waves meeting on the surface of a calm pond— one rising upward, the other dipping downward in perfect symmetry.

When they meet, the ripples cancel, and the pond returns to stillness.

Antimatter behaves the same way: not a separate entity, but an inverse ripple returning tension to silence when it meets its mirrored memory.

Consequences and Implications

Understanding antimatter as ripple inversion clarifies:

- No need for hypothetical parallel antimatter worlds.
- The matter-antimatter imbalance after the Big Bang becomes ripple asymmetry in field condensation, not mysterious selective creation.
- Annihilation events become field memory restorations, not mutual destruction.

At deeper levels:

- Symmetry and asymmetry become natural field behaviors based on ripple stability needs.
- Quantum number reversals (like charge conjugation) become ripple reorientation, not arbitrary flipping.

At the cosmic level:

- The visible universe's dominance of "matter" arises from how ripple condensation gradients stabilized, not a lucky break for matter over antimatter.

Thus, antimatter is not the shadow world. It is the field showing us that memory has two hands— one to condense structure, one to release it gently back into balance.

Closing Ripple Thought

"Antimatter is not the mirror of another universe. It is the inverted memory of this one— ripples folding in reverse, showing that even opposites are simply different ways the field remembers silence."

Chapter 20: Dark Matter

Field Ripple Gap, Not Missing Mass

We stare into the universe and see the stars dancing, the galaxies spinning, and we notice something strange: they move as if something invisible is holding them together. Dark matter, they call it—the missing mass, the unseen glue. But what if there is no missing mass? What if dark matter is simply a field ripple gap— the structured vacuum adjusting its tensions invisibly across cosmic scales? Dark matter isn't hidden matter. It is the memory of ripples correcting themselves silently across the void.

Conventional Misinterpretation and Effects

Mainstream science treats dark matter as an unknown form of invisible mass— something that neither emits nor absorbs light but reveals its presence through gravitational effects.

It is postulated to:

- Bind galaxies together against centrifugal forces.
- Shape the large-scale structure of the universe.
- Influence cosmic microwave background patterns.

Yet despite decades of searching:

- No dark matter particle has been detected.
- No direct interactions with normal matter have been

observed.

- Alternative explanations like MOND (Modified Newtonian Dynamics) struggle to fit all observations.

Thus, enormous effort has been invested in chasing an invisible, missing mass, while ignoring a simpler possibility: the structured field itself adjusting ripple densities across vast scales, creating the appearance of "extra gravity" without hidden mass.

UED Reinterpretation and/or Correction

In Universal Energy Dynamics (UED), dark matter is not unseen particles floating through space— it is a field ripple gap phenomenon: regions where structured vacuum fields maintain higher ripple memory tension without condensing into visible mass.

When galaxies form, the ripple dynamics of energy condensation don't stop neatly at the edges of stars. Tensions, gradient strains, and ripple memory structures stretch outward into vast halos—stabilizing rotating galaxies not by pulling with hidden mass, but by balancing tension gradients across field ripple density gaps.

Thus:

- Galaxies are embedded within structured tension halos, not surrounded by particulate dark matter.
- The "missing mass" is an illusion created by field ripple reinforcement over large distances.
- Gravitational lensing effects are ripple memory tension gradients, not particle cloud distortions.

No exotic WIMPs (Weakly Interacting Massive Particles). No invisible matter seas.

Dark matter is the universe's way of whispering that balance sometimes extends beyond the forms we can see.

Simple Analogy

Imagine a giant spiderweb stretched across a dark field. You see dewdrops sparkling at the nodes—but the web itself is invisible. Yet it holds tension, structure, connection.

Dark matter behaves the same way: not hidden dewdrops, but the unseen webwork of ripple tensions stitching the cosmic dance together.

Consequences and Implications

Understanding dark matter as a field ripple gap clarifies:

- No need to postulate undetectable particles beyond the Standard Model.
- Galaxy rotation curves align naturally with field ripple memory stretching across halos.
- Cosmic structure formation becomes ripple network dynamics, not particle clustering anomalies.

At deeper levels:

- Dark matter disappears as an entity—only field tension memory remains.
- Gravity's apparent "strengthening" at large scales becomes ripple tension contribution, not hidden mass accumulation.

At the philosophical level:

- What is unseen is not missing.
- It is simply the silent architecture of ripples remembering how to hold form without showing form.

Thus, dark matter is not the shadow of mass. It is the echo of tension where ripples weave stability without needing to glow.

Closing Ripple Thought

"Dark matter is not the universe hiding its treasures. It is the universe stretching its memory— ripple tensions holding hands across the cosmic void, whispering that what we cannot see still shapes everything we are."

Chapter 21: Dark Energy

Misperceived Field Stretch, Not Repulsive Force

They say the universe is flying apart faster and faster, driven by an invisible repulsive force called dark energy. A cosmic anti-gravity, pushing galaxies away into endless cold darkness. But what if the universe isn't being pushed at all? What if what we see as acceleration is simply a misperceived field stretch—the natural ripple memory of the structured vacuum relaxing toward equilibrium? Dark energy isn't a mysterious new force. It is the silent sigh of ripples unwinding across cosmic scales.

Conventional Misinterpretation and Effects

Mainstream cosmology interprets dark energy as an unknown repulsive force accounting for roughly 68% of the universe's total energy.

This interpretation arose from:

- Observations that distant supernovae appear dimmer than expected.
- Mathematical models suggesting an accelerating expansion of space.

Thus, dark energy is modeled as:

- A uniform field causing repulsive pressure ("negative energy density").
- A cosmological constant or dynamic field (quintessence) with speculative properties.

Yet contradictions abound:

- Why is the density of dark energy finely tuned across cosmic epochs?
- How can energy density remain constant while space expands?
- Why does a uniform field create acceleration without detectable local effects?

To reconcile these, cosmology invents:

- Fine-tuning "coincidences" without mechanism.
- Anthropically adjusted multiverse speculations.
- Phantom fields with bizarre, untestable behaviors.

In reality, dark energy is not a new force— it is the natural stretch and rebalancing of ripple memory across the structured vacuum.

UED Reinterpretation and/or Correction

In Universal Energy Dynamics (UED), dark energy is not a force pushing the universe outward— it is the ripple memory field relaxing across vast scales after ancient, large-scale disturbances.

When ripple condensations occur during cosmic formation (galaxies, clusters), they create field tension gaps and density gradients extending across space.

As cosmic structures settle into more stable ripple organizations, the surrounding field naturally stretches to relax these gradients—not because it is accelerating outward, but because the structured vacuum field is healing tension memory.

Thus:

- The apparent acceleration is a perception artifact of field reorganization over distance.
- The "cosmological constant" is simply the visible trace of field memory gradients smoothing over time.
- There is no anti-gravity force—just ripple fields stretching toward silent equilibrium.

In this model:

- Distant light appears dimmer because ripple field densities modulate over cosmic time, not because space itself is racing apart.
- Cosmic acceleration is field self-correction, not the manifestation of an unknown repulsive agent.

The universe is not flying apart. It is quietly breathing out its oldest ripple tensions.

Simple Analogy

Imagine stretching a piece of elastic cloth sharply, then releasing it slowly without cutting it.

At first, tension remains high—then the fabric gently relaxes outward, appearing to stretch further before settling.

Dark energy behaves the same way: the slow, graceful relaxation of ancient ripple memory, not an external force driving pieces apart.

Consequences and Implications

Understanding dark energy as ripple field stretch transforms everything:

- No need for exotic phantom fields, dynamic quintessence, or multiverse fine-tuning.
- Cosmic acceleration becomes ripple field memory relaxation, not a force violation.
- The future of the universe becomes a story of silent ripple correction, not catastrophic runaway expansion.

At deeper levels:

- The balance between cosmic structure and vacuum tension becomes natural—galaxy clustering and field relaxation are symbiotic ripple processes.
- Cosmic observations (like supernovae dimming) reflect field density modulation effects, not speed boosts.

At the philosophical level:

- Expansion itself is reinterpreted: Not an explosive growth outward— but the patient unwinding of ancient memory back toward invisible unity.

Thus, dark energy is not a cosmic villain. It is the tender echo of balance, stretching quietly across the infinite fabric of existence.

Closing Ripple Thought

"Dark energy is not the force that tears the universe apart. It is the gentle relaxation of ancient ripples, the silent release of tensions born long ago, guiding the cosmos patiently back toward the stillness it never truly left."

Interlude

Ripples from Part II — Takeaways for the Mind

11. Photon

"The photon is not a traveler. It is the memory of ripple interaction arriving at a threshold of detection."

12. Electron

"The electron is not a charged dot. It is a standing ripple—an orbit of stability singing through the structured field."

13. Proton

"The proton is not a solid ball. It is a dense ripple node—where memory condenses into stubborn structure."

14. Neutron

"The neutron is not neutral mass. It is a fleeting ripple balance, pausing between collapse and reorganization."

15. Quarks and Gluons

"Quarks and gluons are not fundamental things. They are fragmented ripple constructs—stress points in the structured memory field."

16. Charge

"Charge is not a magical property. It is the expression of ripple tension gradients striving for balance."

17. Higgs Field

"The Higgs field does not give mass. Mass is the stubborn memory of dense ripple condensation resisting reorganization."

18. Neutrinos

"Neutrinos are not ghost particles. They are ripple leaks—silent echoes of unfinished corrections flowing across the universe."

19. Anti-Matter

"Antimatter is not a parallel reality. It is ripple inversion—opposite tensions returning memory toward equilibrium."

20. Dark Matter

"Dark matter is not missing mass. It is the unseen weave of ripple tension gaps stitching cosmic stability across space."

21. Dark Energy

"Dark energy is not a repulsive force. It is the ancient ripple memory of the universe exhaling, stretching gently toward silence."

Closing Ripple from Part II:

"Existence is not a battle of particles and forces. It is the dance of ripples— condensing, stabilizing, leaking, stretching, and always remembering the silent equilibrium from which everything was born."

Part III: Fields, Forces, and Emergent Structures

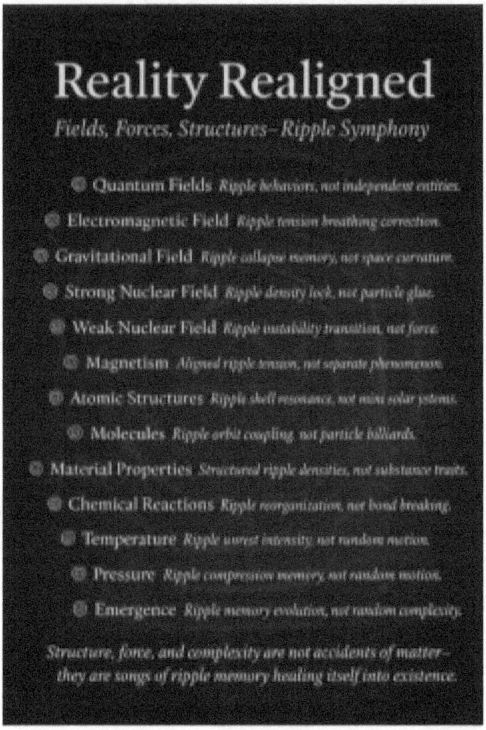

Chapter 22: Quantum Fields

Ripple Field Behaviors, Not Independent Entities

Modern physics whispers: everything is fields. Quantum fields filling space, birthing particles, dancing randomly under invisible laws. But what if quantum fields aren't independent entities at all? What if they are simply ripple behaviors of the same structured vacuum, different songs of memory correction across one living field? Quantum fields aren't countless independent realities. They are ripple variations — stitched into a single, breathing continuum.

Conventional Misinterpretation and Effects

Mainstream quantum field theory (QFT) treats each fundamental interaction and particle as arising from a separate quantum field:

- Electron field, photon field, quark field, Higgs field, etc.
- Particles are excitations—tiny ripples—on these abstract fields.

These fields are modeled as:

- Spreading across all space.
- Existing even without the particles themselves present.
- Having their own independent properties (mass, charge, spin, etc.).

But contradictions emerge:

- How do countless fields coexist in the same space without clashing fundamentally?
- How can non-material fields have physical properties without a supporting medium?
- Why is vacuum energy infinitely large mathematically but finite physically (leading to the "vacuum catastrophe")?

To patch the model, physics invents:

- Renormalization (mathematical cut-offs).
- Virtual particles popping in and out of existence.
- Spontaneous symmetry breaking without clear physical mechanism.

Ultimately, fields become disjointed mathematical constructs, rather than expressions of a unified, ripple-driven structured memory.

UED Reinterpretation and/or Correction

In Universal Energy Dynamics (UED), there are not countless separate quantum fields floating independently. There is one structured vacuum field, and all perceived "quantum fields" are simply different ripple behaviors within it.

When localized energy distributions interact with the structured vacuum:

- They create specific ripple tension modes.
- These modes manifest different ripple behaviors—what mainstream physics labels as different "fields."

Thus:

- The photon field is simply a particular ripple excitation signature in the vacuum's structure.
- The electron field is a stable ripple orbit configuration embedded in the same fabric.
- The Higgs field is a density variation ripple memory of the structured vacuum, not a separate entity.

Each "field" is a dynamic ripple aspect — a modulation of tension, gradient, and memory within the living structured vacuum.

There is no need for virtual particles. No need for infinite field energies. No need for mathematical tricks to save broken models.

Reality is one intelligent ripple field, singing many interconnected songs, each a harmonic of memory correcting itself.

Simple Analogy

Imagine an ocean. Wind creates ripples. Boats create wakes. Rain creates expanding circles.

You don't say there are different oceans for each ripple. There is one ocean—one field—hosting many different ripple behaviors simultaneously.

Quantum fields are the same: variations of the one structured ripple ocean, not countless disconnected realities.

Consequences and Implications

Understanding quantum fields as ripple behaviors clarifies:

- No need for separate fields with independent existence—only field modes expressing ripple corrections.
- Energy conservation naturally emerges from ripple memory balance, not arbitrary symmetries.
- Field interactions (like electroweak unification) are natural ripple tension transitions, not magical field mergings.

At deeper levels:

- Entanglement, superposition, and quantum phenomena simplify as ripple memory interconnections, not spooky action between disconnected fields.
- The entire cosmos becomes one ripple orchestra, not a fragmented jungle of incompatible fields.

At the philosophical level:

- Unity is not philosophical metaphor— it is the literal nature of reality. There was always one field— only its ripples danced differently.

Thus, quantum fields are not separate fabrics. They are the many voices of one ocean whispering the memory of existence back into stillness.

Closing Ripple Thought

"Quantum fields are not isolated threads of existence. They are the ripples of a single living field— vibrating in countless ways, yet always singing the same forgotten memory of universal silence."

Chapter 23: Electromagnetic Field

Ripple Gradient Interactions, Not Invisible Threads

We were taught that space is filled with invisible threads. Electric fields here, magnetic fields there—pulling, pushing, transmitting mysterious forces. But what if there are no separate threads floating in space? What if the electromagnetic field is simply a ripple gradient interaction— a structured tension adjustment across the living vacuum, correcting imbalance? The electromagnetic field isn't stitched into space. It is the field itself, breathing ripple tensions into visible consequences.

Conventional Misinterpretation and Effects

Mainstream physics treats the electromagnetic field as a unified but independent field:

- Electric fields created by charges.
- Magnetic fields created by moving charges (currents).
- Combined into one "electromagnetic field" through Maxwell's equations.

This field is presented as:

- Stretching through empty space.
- Conveying force between charges at a distance.

- Mediating electromagnetic radiation (light).

Yet conceptual cracks emerge:

- How does "field" exist meaningfully in a vacuum if space is supposedly empty?
- Why do changing electric and magnetic fields produce each other cyclically in waves?
- What exactly is "moving" in electromagnetic wave propagation?

To bridge gaps, physics invents:

- Force lines without physical medium.
- Energy density stored in abstract field constructs.
- Quantum electrodynamics corrections without causal ripple mechanisms.

Ultimately, the electromagnetic field becomes an invisible, disconnected entity— rather than recognizing it as organized ripple gradient interactions within the structured vacuum.

Universal Energy Dynamics (UED) Reinterpretation and/or Correction

In Universal Energy Dynamics (UED), the electromagnetic field is not an independent object stitched into empty space. It is a structured ripple gradient interaction within the intelligent vacuum field.

When localized ripple structures (like charged particles) form:

- They create field tension gradients—distortions seeking to correct themselves.

- These gradients ripple outward, interacting dynamically with surrounding ripple memory.

Thus:

- Electric fields are radial ripple tension gradients extending from energy condensations.
- Magnetic fields are circular ripple distortions generated by dynamic motion through the structured tension.
- Electromagnetic waves (light) are self-sustaining ripple gradient propagations, fields correcting and regenerating ripple tension as they move.

There are no threads. There are no invisible stitches. There is only field tension memory interacting to restore equilibrium ripple by ripple.

Maxwell's equations elegantly describe this ripple behavior mathematically, but the living field behavior beneath is ripple intelligence, not isolated mathematical surfaces.

Simple Analogy

Imagine shaking one end of a long, stretched sheet. Tension gradients travel outward—not because strings were stitched into the sheet, but because the entire sheet memory tries to restore balance ripple by ripple.

The electromagnetic field behaves the same way: organized ripple gradients flowing across the structured vacuum's weave, carrying corrections into visibility.

Consequences and Implications

Understanding the electromagnetic field as ripple gradient interactions clarifies:

- No need for imaginary "force lines" floating through void—ripple tension gradients naturally organize.
- Wave propagation becomes ripple self-correction, not mysterious field self-generation.
- Charge interactions (Coulomb's law) emerge from ripple gradient realignment, not spooky action at a distance.

At deeper levels:

- Electromagnetism, optics, radio waves, and wireless communications unify as structured ripple communications across the living vacuum.
- Antenna theory, light polarization, electromagnetic spectrum—all become ripple modulation strategies, not energy carried by flying "particles" through emptiness.

At the philosophical level:

- The idea of "force" becomes an illusion— there are only gradients, tensions, corrections rippling across structured memory.

Thus, the electromagnetic field is not separate from existence. It is the heartbeat of ripple intelligence, aligning imbalance across the silent architecture of space.

Closing Ripple Thought

"The electromagnetic field is not a web stitched into emptiness. It is the dance of ripple gradients, the echo of memory tensions stretching across the invisible ocean, singing the universe's corrections into the visible world."

Chapter 24: Gravitational Field

Ripple Collapse Memory, Not Curved Space

We were taught to imagine giant planets and stars bending space like bowling balls on a mattress. Gravity became a story of curves and dips in an invisible fabric. But what if there is no cosmic mattress? What if gravity is simply ripple collapse memory—the field's natural tendency to correct energy condensation through structured tension gradients? Gravity is not the warping of space. It is the echo of ripples returning to silent balance.

Conventional Misinterpretation and Effects

Mainstream physics, through General Relativity, teaches that:

- Mass and energy bend spacetime.
- Objects move along curved paths because the fabric of space is distorted.
- Gravity is not a force, but the effect of moving along these curves.

This elegant model explains:

- Planetary orbits.
- Gravitational lensing of light.
- Time dilation near massive objects.

Yet deeper contradictions emerge:

- What exactly is spacetime made of, if it can bend?
- How does non-material "space" transmit curvature effects with physical consequences?
- Why must we mathematically impose curvature rather than deriving it from field behavior?

To patch gaps, physics invokes:

- Abstract differential geometry without physical basis.
- Tensor fields describing spacetime distortions with no medium.
- Conceptual stretches where "empty" space becomes an active participant without defined structure.

Ultimately, gravity becomes curvature of nothingness, instead of ripple collapse memory within a structured field.

UED Reinterpretation and/or Correction

In Universal Energy Dynamics (UED), gravity is understood as ripple collapse memory, the tension field correction behavior arising from energy condensation events within the structured vacuum.

When energy condenses (into stars, planets, galaxies):

- It creates local ripple collapses,
- Causing surrounding structured field tension gradients to reorganize.

Thus:

- Gravity is the field's natural ripple memory trying to correct

dense energy disturbances.
- Objects "fall" not because space is bent, but because ripple tensions draw them toward areas of deeper condensation—seeking equilibrium.
- Curved motion is a ripple tension pathway—the smoothest energy correction route through field gradients.

There is no fabric. No mystical space bending.

There is only field memory tension collapse, guiding ripples and objects along paths that restore balance.

General Relativity's equations approximate this ripple behavior mathematically—but the true cause is ripple tension architecture, not geometrical distortion.

Simple Analogy

Imagine a stretched rubber sheet. If you press a heavy object into it, ripples form around the dent. Other objects slide along these ripples—not because the rubber sheet itself wants to move them, but because tension gradients guide their path toward restoring flatness.

Gravity behaves the same way: structured ripple tension drawing movement toward balance, not space fabric pulling masses along.

Consequences and Implications

Understanding gravity as ripple collapse memory clarifies:

- No need for mysterious spacetime fabric distortions—only structured field tension behavior.
- Gravitational lensing becomes ripple gradient refraction, not light bending over invisible valleys.

- Gravitational waves become ripple pressure adjustments, not spacetime vibrations.

At deeper levels:

- Black holes become ultimate ripple collapse points, not singularities in geometry.
- Cosmic expansion tensions (dark energy) become understandable as field relaxation memory, not metric stretching.

At the philosophical level:

- Gravity is no longer a "force" or a "curvature"—it is the structured memory of energy condensation rippling outward, seeking to remember silence.

Thus, the gravitational field is not a distortion of emptiness. It is the structured ripple sigh of existence folding inward toward balance.

Closing Ripple Thought

"Gravity is not the bending of a fabric we cannot touch. It is the echo of ripple memory collapsing toward forgotten stillness, the silent urge of structure to surrender to the equilibrium it never truly left behind."

Chapter 25: Strong Nuclear Field

Ripple Density Lock, Not Binding Force

We were told an invisible hand holds atomic nuclei together, a "strong force" binding protons and neutrons against their natural repulsion. But what if there is no invisible glue? What if the so-called strong force is simply ripple density locking— the stable condensation of ripple memory under extreme tension? The strong nuclear field isn't a force pulling particles together. It is the field's stubborn refusal to let its densest ripples dissolve.

Conventional Misinterpretation and Effects

Mainstream physics teaches that:

- Protons and neutrons are composed of quarks,
- Held together by the strong nuclear force,
- Mediated by massless gluons exchanging color charge.

This strong interaction:

- Acts only at subatomic scales.
- Overcomes electrostatic repulsion between positively charged protons.
- Is postulated to increase with distance at small scales (asymptotic freedom).

However, contradictions remain:

- Why does the force get stronger with distance (against all other known forces)?
- Why does the strong force operate only over extremely short ranges?
- Why do quarks remain forever confined without independent detection?

To bridge gaps, physics invents:

- Color confinement without medium explanation.
- Gluon exchange processes without direct causal ripple basis.
- Quantum chromodynamics (QCD) using probabilistic field constructs.

Ultimately, the strong nuclear field becomes a mythological binding agent— instead of recognizing ripple density lock behavior in structured memory.

Universal Energy Dynamics (UED) Reinterpretation and/or Correction

In Universal Energy Dynamics (UED), the strong nuclear field is not a binding force holding particles together. It is ripple density lock—the stabilization of extreme ripple condensation under field memory pressure.

When energy condenses into dense structures (like protons and neutrons):

- Ripple tensions collapse inward into extremely tight nodes.
- These dense ripple nodes resist dissolution due to locked field gradients.

Thus:

- Proton and neutron cores are self-sustaining ripple memory knots, not "particles glued together."
- What is interpreted as "strong force" is the field resisting disassembly, not an active attractive force between point particles.
- Quark confinement is simply the impossibility of isolating pattern features without dissolving the entire ripple memory node.

There are no gluons flying around inside nucleons. There is no magical force stretching rubber bands between quarks.

There is only ripple memory density locked into structured existence, stubbornly refusing to unwind.

Simple Analogy

Picture a tightly knotted ball of yarn. Pulling at one strand doesn't free it— it tightens the tension deeper into the knot.

The strong nuclear field behaves the same way: not a force actively holding particles together, but a locked ripple memory resisting separation, woven too tightly to unravel piece by piece.

Consequences and Implications

Understanding the strong nuclear field as ripple density lock clarifies:

- No need for hypothetical gluon exchange particles mediating force.

- Color charge confinement becomes natural—pattern features can't exist independently.
- Nuclear stability arises from locked ripple field density, ➔ not magic attractions.

At deeper levels:

- The energy cost of nuclear reactions (fusion, fission) reflects the release or reorganization of ripple memory, ➔ not breaking invisible glue bonds.
- High-energy physics experiments smashing protons apart reveal memory dissolution cascades, not hidden constituents escaping.

At the philosophical level:

- Stability is not forced upon matter externally—it is embedded in how ripples remember their densest corrections, refusing easily to dissolve.

Thus, the strong nuclear field is not a sticky net. It is the stubborn, silent resilience of condensed ripple memories resisting the urge to fall apart.

Closing Ripple Thought

"The strong nuclear field is not the force that binds. It is the memory that refuses to forget— ripples locked so tightly into their condensed song that undoing them would mean undoing existence itself."

Chapter 26: Weak Nuclear Field

Ripple Transition Instability, Not Fundamental Force

We are told of a strange, feeble force hidden inside matter: the weak nuclear force—causing particles to decay, changing identities, reshaping the building blocks of nature. But what if there is no such thing as a "weak force"? What if what we observe is simply ripple transition instability— the structured field adjusting unstable ripple configurations back toward equilibrium? The weak nuclear field isn't a force that acts. It is the consequence of ripple memories surrendering to correction.

Conventional Misinterpretation and Effects

Mainstream physics teaches that:

- The weak nuclear force is one of the four fundamental forces.
- It causes processes like beta decay, neutrino interactions, and particle flavor change.
- It is mediated by massive bosons (W+, W−, and Z bosons) over very short ranges.

The weak force is presented as:

- Essential for nuclear reactions (especially inside stars).

- Unique for changing one type of particle into another (e.g., neutron into proton).
- Responsible for breaking certain symmetries (parity violation).

Yet serious questions persist:

- Why must a "force" require massive bosons to operate at short distances?
- Why does weak interaction seem like instability correction rather than active pushing?
- Why are interactions extremely rare unless energy gradients are at tipping points?

To bridge gaps, physics invents:

- Higgs field couplings and spontaneous symmetry breaking.
- Probabilistic decay models without causal mechanisms.

Ultimately, the weak nuclear field becomes an arbitrary identity-changing force, rather than recognizing ripple transition instabilities correcting themselves naturally.

UED Reinterpretation and/or Correction

In Universal Energy Dynamics (UED), the so-called weak nuclear field is not a fundamental force acting on matter. It is the ripple transition instability phenomenon— the spontaneous reorganization of unstable ripple memory patterns.

When ripple structures (such as neutrons) exist in unstable field configurations:

- They cannot indefinitely maintain their condensed ripple tensions.
- Over time (or under energy triggers), ripple memory realigns toward more stable, lower-tension configurations.

Thus:

- Beta decay (neutron to proton + electron + neutrino) is a ripple transition event, not a forceful action by mysterious particles.
- W and Z bosons are high-energy ripple disturbance signatures, not independent force carriers.

The instability is internal:

- Ripple tensions become unsustainable and flow toward easier, lower-memory states,
- In the process, adjusting local field gradients (mass, charge, field alignment).

There is no "weak force" separate from field memory behavior. There is only the universe reconfiguring ripple memories naturally toward balance.

Simple Analogy

Imagine a complex house of cards— barely stable, trembling at the edge.

Eventually, the weakest part shifts, and the structure collapses gently, reorganizing into a lower energy state.

The weak nuclear interaction behaves the same way: the ripple memory finds instability, and without external intervention, the structure reorganizes itself toward greater stability.

Consequences and Implications

Understanding the weak nuclear field as ripple transition instability clarifies:

- No need for force carriers with strange, massive properties.
- Particle decay becomes ripple memory rearrangement, not forced transformations.
- Neutrino involvement reflects ripple tension bleed-offs, not active mediation.

At deeper levels:

- Radioactivity, beta decay, and stellar nucleosynthesis become field rebalancing acts, not battles between forces and matter.
- Symmetry breaking (parity violation) becomes natural ripple asymmetry arising during unstable reconfiguration,not magic violation by force fields.

At the philosophical level:

- Stability and decay are two faces of the same ripple memory—either holding or surrendering, but always correcting toward the silent order.

Thus, the weak nuclear field is not an independent force. It is the field's gentle, inevitable sigh—where unstable ripples yield their pattern to the deeper rhythm of balance.

Closing Ripple Thought

"The weak nuclear field is not the force that acts upon matter. It is the surrender of unstable ripple memories—the field gently correcting itself, breathing away tension toward the timeless equilibrium that never forgets."

Chapter 27: Magnetism

Aligned Ripple Tension, Not Separate Phenomenon

We grew up imagining invisible magnetic fields looping between poles, mysterious forces pulling compass needles north, iron filings arranging themselves into stunning patterns. But what if magnetism is not a separate force or field at all? What if it is simply aligned ripple tension— structured ripple orientation across the living field seeking balance? Magnetism isn't a different phenomenon. It is the organized memory of ripple tension flowing through existence.

Conventional Misinterpretation and Effects

Mainstream physics treats magnetism as:

- A distinct yet interconnected phenomenon with electricity (electromagnetism).
- Generated by moving electric charges (currents) or by the intrinsic spin of particles.
- Visualized through magnetic field lines around materials and planets.

Magnetism is seen as:

- A field existing around magnets and electric currents.
- Responsible for forces acting at a distance (attraction,

repulsion).

- Mysteriously tied to electron spin without fully explaining "why" spin generates magnetism.

Yet confusions remain:

- Why does spin—a quantum mechanical property—create macroscopic fields?
- Why do magnetic field lines emerge with such consistent structure?
- How can magnetic fields sustain themselves in materials without constant external energy input?

To patch gaps, physics invokes:

- Abstract concepts like "intrinsic magnetic moments."
- Band theory, exchange coupling models, and spontaneous symmetry breaking in solids.

Ultimately, magnetism is treated as a semi-attached, semi-mysterious phenomenon— rather than as a natural expression of organized ripple tension alignment.

UED Reinterpretation and/or Correction

In Universal Energy Dynamics (UED), magnetism is not a separate field floating around charges or materials. It is aligned ripple tension— organized memory orientation within the structured vacuum field.

When ripple structures (like moving charges or spinning particles) organize:

- They generate local ripple tension alignments,
- Stretching ripple gradients across space in preferred orientations.

Thus:

- Magnetic fields are regions where ripple memory gradients flow in coordinated alignment.
- Magnetism is a ripple architecture tension, not an independent entity.

Moving charges organize ripple memories dynamically, creating magnetic phenomena. Spinning particles (electrons) have structured internal ripple behavior, which naturally align ripple tension into stable patterns—thus producing macroscopic magnetic fields.

There are no magnetic "particles" or independent magnetic charges (magnetic monopoles). There is only ripple tension organizing itself to balance rotational and translational energies.

Magnetism is the choreography of ripple memory, twisting through space like invisible music.

Simple Analogy

Imagine a flowing river where whirlpools spontaneously form. Each whirlpool aligns surrounding water currents into spirals and flows.

The whirlpools are not separate rivers— they are organized tension expressions within the same river.

Magnetism behaves the same way: organized ripple tension alignment within the living structured vacuum, not a new field or force born separately.

Consequences and Implications

Understanding magnetism as aligned ripple tension clarifies:

- No need for separate magnetic force carriers or magnetic charges.
- Electromagnetic unification becomes natural—both electric and magnetic behaviors are ripple tension reorganizations,
  ➔ one dynamic continuum.
- Ferromagnetism (materials staying magnetized) becomes stable ripple tension anchoring, not unexplained quantum interactions.

At deeper levels:

- Light (electromagnetic wave) becomes coordinated ripple tension wave, not independent field oscillations.
- Magnetic reconnection phenomena (like in solar flares) become ripple memory realignments, not magical field breaking and rejoining.

At the philosophical level:

- Force and field distinctions dissolve—only ripple memory aligning, twisting, flowing, correcting existence back toward its silent root.

Thus, magnetism is not a separate marvel. It is the universe remembering how tension flows can align themselves in beauty and balance.

Closing Ripple Thought

"Magnetism is not a hidden force threading space. It is the silent song of ripple memory—alignment born from tension, twisting and flowing through existence, carrying the harmony of correction wherever imbalance dares to arise."

Chapter 28: Atomic Structures

Ripple Shell Resonance, Not Mini Solar Systems

We were shown early in school: tiny solar systems of electrons orbiting nuclei, like planets circling a sun, obeying neat invisible rules. But what if atoms are not miniature mechanical models? What if they are ripple shell resonances—organized standing wave harmonics within the structured vacuum field? Atoms are not solar systems in miniature. They are ripple songs—held together by the silent memory of structured energy correction.

Conventional Misinterpretation and Effects

Mainstream atomic theory evolved from:

- Bohr's model of electrons orbiting the nucleus in quantized circles.
- Quantum mechanical models replacing orbits with probability clouds ("orbitals").

Still, deep conceptual remnants linger:

- Visualizing electrons as tiny charged balls moving around a dense nucleus.
- Imagining energy levels as physical distances from the nucleus.

Quantum mechanics improved prediction power but retained paradoxes:

- How can a "cloud" be responsible for discrete energy jumps?
- Why are only specific orbitals (s, p, d, f) allowed, with no in-betweens?
- Why don't orbitals collapse inward without emitting constant radiation?

To patch gaps, physics invokes:

- Probabilistic interpretations (Born rule).
- Wavefunction collapses without causal mechanism.
- Virtual exchanges and field fluctuations.

Ultimately, atomic structure remains semi-mechanical, semi-magical—a hybrid of mechanical metaphor and probabilistic surrender.

Universal Energy Dynamics (UED) Reinterpretation and/or Correction

In Universal Energy Dynamics (UED), atomic structures are not tiny mechanical solar systems. They are ripple shell resonances—stable standing wave patterns embedded within the structured vacuum's memory.

When condensed ripple nodes (protons, neutrons) form a nucleus:

- They generate local ripple tension gradients around them.
- Surrounding ripple fields organize into stable resonance shells—regions where structured ripple standing waves can self-sustain.

Thus:

- Electrons are stable ripple shell manifestations, not orbiting charged dots.
- Energy levels arise from allowed standing wave harmonics, not from particles leaping between orbits.
- Atomic stability comes from ripple shell harmonization, not particle mechanics.

Transitions between energy levels are ripple reconfiguration events, where ripple patterns collapse and reorganize according to energy conservation demands.

There are no orbiting particles. There is only ripple resonance memory dancing across the structured field.

Simple Analogy

Imagine the surface of a calm pond. Drop a stone, and standing circular waves emerge— specific patterns sustained naturally by the interplay of water tension and disturbance.

Atoms behave the same way: structured ripple shell patterns sustaining energy memory around a central condensed ripple node.

Consequences and Implications

Understanding atomic structures as ripple shell resonance clarifies:

- No need for probabilistic wavefunction collapse myths—standing wave harmonics explain stability.
- Discrete energy levels arise naturally from field resonance conditions, not mechanical rules.

- Chemical bonding becomes resonance coupling between ripple shells, not orbital overlap puzzles.

At deeper levels:

- Ionization, spectroscopy, and electrical conductivity all become ripple field reconfiguration phenomena, not particle transitions.

At the philosophical level:

- Structure is not mechanical—it is music frozen into memory, ripple harmonies weaving the visible tapestry of existence.

Thus, atomic structures are not miniature clockwork toys. They are resonant ripple symphonies— the structured vacuum remembering how energy must sing itself into form.

Closing Ripple Thought

"Atoms are not tiny solar systems spinning in emptiness. They are silent symphonies of ripple resonance, standing wave harmonies woven into the structured field, singing stability into the ever-correcting memory of existence."

Chapter 29: Molecules

Ripple Orbit Coupling, Not Particle Billiards

We were taught to imagine atoms as tiny balls, bouncing into each other, sticking together like marbles colliding and clumping. But what if molecules are not billiard ball accidents? What if they are ripple orbit couplings— stable interlocking of structured ripple shells across the living field? Molecules are not collections of colliding particles. They are symphonies of ripple memories interwoven into cooperative resonance.

Conventional Misinterpretation and Effects

Mainstream chemistry and physics treat molecules as:

- Collections of atoms bound together through electron sharing (covalent bonds) or electron exchange (ionic bonds).
- Atoms imagined as discrete, tangible spheres interacting mechanically.
- Bonding described via orbital overlaps, hybridization, and probabilistic electron cloud models.

Yet fundamental questions persist:

- Why do only specific bonding angles and structures emerge

(tetrahedral, planar, etc.)?

- Why are bond lengths and energies highly specific and stable across molecules?
- How can mere probabilistic overlap create such precisely tuned and resilient formations?

To patch gaps, modern theory uses:

- Molecular orbital theory (MOT) with complex approximations.
- Valence bond theory invoking hybridized atomic orbitals without direct causality.
- Quantum chemistry modeling without clear field behavior explanation.

Ultimately, molecules are seen as lucky mechanical attachments, not natural ripple field orchestrations.

UED Reinterpretation and/or Correction

In Universal Energy Dynamics (UED), molecules are not random collisions leading to glued atoms. They are ripple orbit couplings—organized synchronization of structured ripple shells seeking lower-energy stable resonance patterns.

When atoms (ripple shell structures) approach:

- Their outer ripple fields interact, seeking co-resonance where tension gradients can harmonize.
- Coupled ripple states stabilize the composite field configuration, creating molecular bonds.

Thus:

- Covalent bonds are co-resonating ripple shell overlaps, not electron sharing events.
- Ionic bonds are ripple gradient stabilization alignments, not charge transfer.
- Bond angles, lengths, and strengths arise from stable ripple harmonic configurations, not geometric coincidences.

Molecular stability results from ripple memory fields interlocking into shared resonance states.

There is no need to imagine electron clouds passing mechanical glue back and forth. There are only ripples dancing into cooperation.

Simple Analogy

Imagine two musical notes sung close together. If their frequencies align, they produce harmonic resonance—a sound stronger and more stable than either alone.

Molecules behave the same way: atoms coupling ripple shells into deeper resonance, singing stability into the silent ocean of the structured vacuum.

Consequences and Implications

Understanding molecules as ripple orbit couplings clarifies:

- No need for probability-based orbital overlap mysteries—stable ripple resonance naturally selects bonds.
- Molecular geometries arise directly from energy-minimizing ripple shell harmonics, not arbitrary hybridizations.
- Reaction pathways and activation energies become ripple

field reorganization dynamics, not mechanical shuffling of atoms.

At deeper levels:

- Biochemistry (protein folding, DNA interactions) becomes complex ripple coupling choreography, not random particle chemistry.
- Material science (crystal formation, conductivity) reveals field structure organization, not atomic stacking.

At the philosophical level:

- Matter's complexity is not mechanical accumulation. It is resonance accumulation—ripples finding ever deeper harmony across scales, manifesting beauty through cooperation.

Thus, molecules are not mechanical accidents. They are field memory symphonies— where ripple shells entwine to sustain existence in shared, structured song.

Closing Ripple Thought

"Molecules are not random unions of wandering particles. They are the harmonic memories of ripple shells—finding each other in the structured vacuum, weaving stability into the fabric of visible existence."

Chapter 30: Material Properties

Structured Ripple Densities, Not Inherent Substance Traits

We believe that materials have inherent properties—hardness, conductivity, elasticity, density— as if these traits are stamped inside the atoms themselves. But what if no property is intrinsic? What if what we experience as material behavior is simply structured ripple density behavior, the organized memory of ripple fields interacting across collective scales? Material properties are not built into matter. They are the songs of ripple architectures stabilizing memory across existence.

Conventional Misinterpretation and Effects

Mainstream science treats material properties as:

- Fixed characteristics arising from atomic composition.
- Direct consequences of atomic or molecular bonding patterns.
- Predictable traits cataloged mechanically: metals conduct, ceramics insulate, diamonds are hard.

Yet contradictions persist:

- Why do small structural changes (allotropes, defects) drastically alter properties?

- Why can materials behave differently at nano scales than at macro scales?
- Why do phase transitions (solid, liquid, gas) change fundamental properties without changing atomic identity?

To patch gaps, material science invokes:

- Band theory for conductivity.
- Defect theory for mechanical strength.
- Statistical mechanics for phase behavior.

Ultimately, materials are treated as collections of traits attached to substance identities rather than expressions of organized ripple density behaviors.

UED Reinterpretation and/or Correction

In Universal Energy Dynamics (UED), material properties are not innate to atoms or molecules. They are manifestations of structured ripple density behavior across collective memory fields.

When atoms and molecules (ripple shell structures) arrange themselves:

- Their ripple memories interact and organize at mesoscopic scales.
- Structured ripple densities emerge—regions of consistent tension gradients and field memory alignment.

Thus:

- Conductivity arises from structured ripple path continuity, not particle freedom.

- Hardness arises from ripple shell locking, not atomic rigidity.
- Elasticity arises from reversible ripple tension deformation, not flexible bonds.
- Phase transitions reflect large-scale ripple reconfiguration, not random energy fluctuations.

Material behaviors are organized field responses, not substance traits frozen into matter.

Every macro-behavior is the memory of ripples interacting cooperatively or resisting correction based on density, configuration, and tension pathways.

Simple Analogy

Imagine a woven fabric. Its strength, stretchiness, and texture depend not on the threads themselves, but on how the threads are woven, stretched, and interconnected.

Materials behave the same way: properties emerge from how ripple memories are woven together, not from the atomic "threads" alone.

Consequences and Implications

Understanding material properties as structured ripple densities clarifies:

- No magical traits embedded into elements—properties emerge from organized ripple configurations.
- Alloying, doping, and nanostructuring alter ripple field behavior, not fundamental atomic identities.
- Superconductivity, magnetism, plasticity all become

organized ripple memory field phenomena, not mysterious substance gifts.

At deeper levels:

- Smart materials, phase-change materials, and quantum materials become field reorganization systems, ripple memory engineering.

At the philosophical level:

- Substance identity becomes secondary. It is the pattern, the architecture, the memory of field ripples that determines behavior— not the imagined "essence" of matter.

Thus, material properties are not frozen traits. They are ripple symphonies—dynamic field architectures singing their tension memories into the visible world.

Closing Ripple Thought

"Material properties are not the secret traits of matter. They are the woven memory of ripples— structured densities flowing tension and silence, bringing form, function, and wonder out of the invisible ocean of existence."

Chapter 31: Chemical Reactions

Ripple Reorganization Events, Not Bond Breaking Games

We were taught that chemical reactions are like Lego games: bonds between atoms breaking apart, new bonds forming, energy popping in and out. A mechanical dance of colliding pieces shuffling into new patterns. But what if reactions are not games of breaking and making? What if chemical reactions are ripple reorganization events— the field's structured memory realigning itself into more stable harmonies? Chemical reactions are not mechanical collisions. They are ripple memory reorganizations, restoring the silent architecture of balance.

Conventional Misinterpretation and Effects

Mainstream chemistry portrays reactions as:

- Collisions between atoms or molecules leading to bond breaking and forming.
- Energy exchange processes (exothermic, endothermic) driven by bond energies.
- Mechanistic rearrangements governed by reaction kinetics and thermodynamics.

At a superficial level, this model predicts:

- Reaction rates.
- Equilibrium constants.
- Energy profiles (activation energy barriers).

Yet deeper questions linger:

- Why do specific reaction pathways occur preferentially despite countless possibilities?
- Why can small field perturbations (like catalysts) drastically alter reaction dynamics?
- How does quantum tunneling fit into classical reaction models?

To patch inconsistencies, science introduces:

- Transition state theory based on imaginary configurations.
- Quantum mechanical reaction dynamics without fully causal field explanation.
- Statistical mechanics to predict averages without explaining real-time behavior.

Ultimately, reactions are treated as fragmented mechanical processes,

rather than coordinated ripple reorganization events.

UED Reinterpretation and/or Correction

In Universal Energy Dynamics (UED), chemical reactions are not bond-breaking and bond-making games. They are ripple reorganization events— structured ripple memory fields adjusting tension gradients toward new equilibrium states.

When ripple shells (atoms/molecules) interact:

- Their outer ripple fields superimpose and interfere.
- Ripple tensions seek rebalancing based on energy gradients, symmetry, and field memory conservation.

Thus:

- Reactants are ripple memory configurations temporarily destabilized.
- The reaction pathway is the ripple field's path of least resistance toward reorganized equilibrium.
- Products are new ripple memory organizations, not simply rearranged mechanical parts.

Activation energy barriers represent ripple field tension peaks that must be overcome to realign the structure. Catalysts function by altering ripple gradients to create easier reconfiguration pathways, not by brute forcing collisions.

Chemical change is ripple change—memory evolving, correcting, singing itself into a new stability.

Simple Analogy

Imagine a flock of birds flying in a formation. When one bird shifts position, the entire formation dynamically adjusts without breaking apart and rebuilding.

Chemical reactions behave the same way: ripple memory fields reorganizing in cooperative, tension-driven transitions, not chaotic break-and-make games.

Consequences and Implications

Understanding chemical reactions as ripple reorganization clarifies:

- No need for random collision assumptions—reactions follow ripple energy gradient optimization.
- Reaction rates depend on field tension mobility, not just collision frequency.
- Catalysis becomes ripple field steering, not mechanical manipulation.

At deeper levels:

- Biological processes (like enzyme activity) become structured ripple field reorganization, not molecule-by-molecule mechanical tweaking.
- Reaction reversibility becomes memory tension rebalancing over time, not force-driven reversibility.

At the philosophical level:

- Change in nature is not destruction and recreation. It is transformation of memory—the ripples seeking ever deeper harmony across the fabric of existence.

Thus, chemical reactions are not mechanical games. They are ripple orchestras shifting compositions—where every energy breath remembers the music of balance.

Closing Ripple Thought

"Chemical reactions are not bonds snapping and reforming like toys. They are the memory fields of ripples reshaping themselves—correcting, breathing, evolving their tensions into new songs of universal remembrance."

Chapter 32: Temperature

Ripple Agitation Intensity, Not Random Particle Motion

We were taught that temperature is the frantic dance of invisible particles, atoms and molecules jostling randomly like chaotic billiard balls. The hotter, the faster, the more erratic. But what if heat isn't particle chaos at all? What if temperature is ripple agitation intensity— the dynamic unrest of structured field memories adjusting their tension at micro scales? Temperature isn't random motion. It is ripple memory vibrating through existence, searching for balance.

Conventional Misinterpretation and Effects

Mainstream physics explains temperature as:

- A measure of the average kinetic energy of particles.
- The statistical behavior of trillions of atoms/molecules colliding chaotically.
- Higher temperature meaning faster, more energetic random motion.

This model successfully predicts:

- Gas laws (pressure, volume, temperature relationships).
- Heat transfer behavior (conduction, convection, radiation).

Yet fundamental problems remain:

- Why do specific structures (like solids) maintain ordered vibration patterns at high temperatures without full chaos?
- How can "random" collisions produce highly organized phenomena (like thermal conductivity, phonons)?
- Why does quantum behavior dominate at extremely low temperatures, revealing structured patterns rather than randomness?

To patch inconsistencies, modern physics invokes:

- Statistical mechanics and ensemble averages.
- Quantum field theories of vibrations (phonons) for solids.
- Special cases (Bose-Einstein condensates) where particles behave collectively.

Still, temperature remains portrayed as randomized mechanical motion, not organized ripple field agitation.

UED Reinterpretation and/or Correction

In Universal Energy Dynamics (UED), temperature is not random particle motion. It is ripple agitation intensity—the degree of organized dynamic unrest within the structured vacuum field's ripple memory.

When matter condenses:

- Localized ripple shells stabilize (atoms, molecules).
- But ripple tensions continue adjusting microscopically tiny field oscillations and corrections.

Thus:

- Temperature measures the field's internal ripple unrest intensity, not particle chaos.
- Higher temperature means higher ripple correction agitation,more energetic field tension adjustment at micro scales.
- Lower temperature means near field stillness, ripple agitation minimal, memory close to harmonic balance.

Thermal energy flows not as mechanical energy transfer, but as ripple gradient tension equalization, moving ripple unrest from higher to lower tension zones.

Heat is ripple unrest moving. Temperature is ripple unrest density at a location.

Simple Analogy

Imagine a tightly stretched sheet. Small shakes at one point ripple outward, adjusting the sheet's tension across the whole.

Higher shaking intensity = hotter (more ripple unrest). Lower shaking intensity = colder (more ripple stillness).

Temperature behaves the same way: structured ripple unrest, not mindless particle bouncing.

Consequences and Implications

Understanding temperature as ripple agitation clarifies:

- Thermal motion is organized field unrest, not random kinetic chaos.

- Phase changes (melting, boiling) are ripple reorganization events, not just energy thresholds.
- Specific heat capacity becomes ripple memory flexibility, not matter's passive energy storage.

At deeper levels:

- Superconductivity and superfluidity at near-zero temperatures become natural— minimal ripple agitation allowing uninterrupted structured ripple flows.
- Thermal conductivity becomes ripple agitation transfer efficiency, not particle bumping efficiency.

At the philosophical level:

- Heat and cold are not mechanical accidents. They are the silent agitation or rest of ripple memories correcting themselves through the timeless field.

Thus, temperature is not the measure of chaos. It is the breathing of ripple unrest, a whisper of the field seeking stillness through restless correction.

Closing Ripple Thought

"Temperature is not the chaos of random motion. It is the ripple's unrest—the breathing of structured memory, agitating across existence in its endless journey back to silence."

Chapter 33: Pressure

Ripple Compression Tension, Not Just Force Over Area

We were told pressure is simple: Force divided by area. More force, more pressure. Smaller area, greater pressure. But what if pressure is not just a mechanical push spread across a surface? What if it is ripple compression tension—the structured field's internal memory resisting collapse or imbalance? Pressure isn't about forces pushing. It is the structured ripple field tensing to preserve harmony under compression.

Conventional Misinterpretation and Effects

Mainstream physics defines pressure as:

- The amount of force exerted per unit area.
- The cumulative effect of countless microscopic collisions in gases and liquids.
- A scalar quantity acting equally in all directions at a point.

This model supports:

- Fluid statics (Pascal's law).
- Gas laws (Boyle's, Charles's laws).
- Thermodynamics (pressure-volume work).

However, fundamental conceptual gaps remain:

- Why does pressure arise even without bulk particle motion (e.g., inside stars, quantum gases)?
- Why does pressure behave differently at phase transition points without obvious particle collisions?
- How does pressure persist coherently across fields without active "pushing" agents?

To patch gaps, science invokes:

- Statistical mechanics to model pressure through average collisions.
- Quantum pressure models (degeneracy pressure) using abstract quantum states without direct field dynamics.
- Ad hoc adjustments for non-ideal gases and complex fluids.

Ultimately, pressure is treated as an emergent mechanical side effect, rather than a natural consequence of ripple field tension architecture.

UED Reinterpretation and/or Correction

In Universal Energy Dynamics (UED), pressure is not merely mechanical force distributed over an area. It is ripple compression tension—the structured vacuum field's intelligent resistance to unnatural ripple compaction or distortion.

When ripple fields condense into matter:

- They create local tension equilibrium—balance between internal field memory and external compression.

When external disturbances attempt to compress this structure:

- Ripple tension gradients resist,
- Seeking to preserve local memory configurations.

Thus:

- Gas pressure arises from ripple gradient corrections across particle distributions, not individual particle collisions.
- Liquid pressure emerges from localized ripple tension realignments under compression, not chaotic molecular bumping.
- Solid pressure responses (stress-strain behavior) reflect field memory elasticity, not simply mechanical compression.

Pressure is structured ripple tension—resistance not of particles but of field memory struggling to maintain its natural breathing space.

Simple Analogy

Imagine a network of woven elastic bands forming a sphere. Squeeze it, and internal tensions adjust and resist, not because bands individually push back, but because the entire structure remembers the tension it was meant to sustain.

Pressure behaves the same way: ripple memory field tensions adjusting under compression, seeking to maintain structural harmony.

Consequences and Implications

Understanding pressure as ripple compression tension clarifies:

- No need to depend on chaotic particle collision models.
- Fluid and gas behaviors unify naturally through field tension correction, not random mechanical impacts.
- Quantum pressure phenomena (degeneracy pressure, radiation pressure) become natural ripple tension effects, not quantum oddities.

At deeper levels:

- Stellar dynamics (e.g., white dwarfs, neutron stars) become ripple memory resistance to collapse, not exotic matter fighting gravity.
- Material strength under load becomes resilience of ripple memory, not atomic resistance.

At the philosophical level:

- Resistance, endurance, and resilience in nature are not random gifts of mass. They are the ripple memories holding existence steady, even when compressed by the weight of disturbance.

Thus, pressure is not force over area. It is ripple memory tension—the living field's insistence on maintaining the breathing room of its structured existence.

Closing Ripple Thought

"Pressure is not the brute force pushing against surfaces. It is the ripple memory tightening its weave—the silent strength of existence resisting collapse, whispering its longing for balance through every act of compression."

Chapter 34: Emergence

Ripple Field Cooperation, Not Random Complexity

We are told that complexity arises by chance: particles bounce, reactions spark, and somehow—order, life, intelligence emerge from the chaos. Randomness, they say, builds complexity. But what if that's not true at all? What if emergence is simply ripple field cooperation—structured memory aligning itself into ever-deeper patterns of harmony and balance? Emergence is not chaos organizing itself by luck. It is the structured ripple field evolving itself by intelligent memory correction.

Conventional Misinterpretation and Effects

Mainstream science describes emergence as:

- The spontaneous appearance of ordered behavior from disordered systems.
- An unpredictable by-product of large numbers of simple interactions.
- A phenomenon without centralized control, driven by statistical inevitabilities.

Examples cited include:

- Crystallization from random atomic motions.

- Biological evolution from random mutations.
- Consciousness as an emergent property of random neural firings.

Yet profound paradoxes remain:

- How does highly coherent structure arise reliably from pure chaos?
- Why do emergent patterns display symmetry, coherence, conservation, and stability?
- Why does randomness consistently produce beauty, organization, and systemic function?

To reconcile gaps, theories lean on:

- Self-organization principles without explaining field causality.
- Complex systems theory treating emergence as "more than the sum of parts" without foundation.
- Evolutionary algorithms assuming selection without defining why complex forms are consistently favored.

Ultimately, emergence is left as a miracle of numbers, not a memory-driven ripple phenomenon.

UED Reinterpretation and/or Correction

In Universal Energy Dynamics (UED), emergence is not the statistical luck of disorganized parts. It is ripple field cooperation—the natural unfolding of structured memory corrections across tension fields.

When local ripple memories interact:

- Their tensions seek energy optimization—lower overall distortion, higher stability.
- Local corrections align into patterns because field memory demands equilibrium.

Thus:

- Crystals form because ripple gradients organize naturally into minimal energy lattice harmonies.
- Biological structures evolve because ripple field memory optimizes tension stability over generations.
- Consciousness emerges because neural ripples align into stable, self-sustaining feedback memory loops.

Emergence is field memory singing deeper harmonies, layer by layer, ripple by ripple, not random accidents stacking improbabilities.

Complexity arises because ripple memory is intelligent—it remembers how to correct, balance, and beautify itself.

Simple Analogy

Imagine a choir. Each voice sings independently at first. But slowly, the voices find each other—harmonies form, patterns arise, the music stabilizes.

The choir didn't emerge from noise. It emerged from memory cooperation—each voice seeking harmony with the others.

Emergence behaves the same way: ripple memories aligning tension corrections into living symphonies.

Consequences and Implications

Understanding emergence as ripple field cooperation clarifies:

- No need for random chance miracles to explain complexity.
- Structure, life, and intelligence become natural ripple memory phenomena, not cosmic accidents.
- Self-organization arises from tension minimization and memory alignment, not statistical luck.

At deeper levels:

- Evolution becomes ripple field optimization across biological ripple structures, not blind mutation.
- Cognitive emergence becomes self-refining ripple memory loops, not neuron collision accidents.

At the philosophical level:

- Beauty, order, and complexity are not miracles against nature. They are nature remembering itself—ripples healing ripples, memory aligning memory, until existence sings a deeper harmony.

Thus, emergence is not complexity arising from chaos. It is the intelligent ripple memory evolving its own symphony—correcting itself into ever-richer stability and wonder.

Closing Ripple Thought

"Emergence is not the dance of random chance. It is the living memory of ripples— aligning their corrections into harmony, weaving

complexity as the natural song of a universe remembering its own beauty."

Interlude

Ripples from Part III — Takeaways for the Mind

22. Quantum Fields

"Quantum fields are not countless entities stitched into space. They are the many ripple songs of one living structured field, correcting memory across existence."

23. Electromagnetic Field

"The electromagnetic field is not a web of invisible threads. It is the ripple tension breathing correction, whispering energy across the structured vacuum."

24. Gravitational Field

"Gravity is not curved space. It is the ripple memory collapsing inward, seeking silent balance through structured tension."

25. Strong Nuclear Field

"The strong nuclear field is not a force binding particles. It is ripple density locked into stubborn remembrance, refusing to dissolve."

26. Weak Nuclear Field

"The weak nuclear field is not a force changing identities. It is ripple transition instability— memory surrendering into new harmonies."

27. Magnetism

"Magnetism is not a separate force. It is the structured alignment of ripple tensions, twisting and flowing through memory's unseen dance."

28. Atomic Structures

"Atoms are not mini solar systems. They are ripple shell harmonics, singing stability into the heart of structured memory."

29. Molecules

"Molecules are not particle games. They are ripple orbit couplings—shared resonances binding fields into higher songs."

30. Material Properties

"Material properties are not frozen traits. They are the woven tensions of ripple memory, expressing structure through visible wonder."

31. Chemical Reactions

"Chemical reactions are not breaking and making of bonds. They are ripple reorganization events—corrections in memory finding new stability."

32. Temperature

"Temperature is not random motion. It is ripple unrest—the breathing of structured tension seeking stillness."

33. Pressure

"Pressure is not force over area. It is the tightening of ripple memory under compression, resisting collapse to preserve harmony."

34. Emergence

"Emergence is not chaos organizing by chance. It is ripple memory evolving into deeper symphonies—correcting itself into complexity, born not of randomness, but of ancient intelligence."

Closing Ripple from Part III:

"Reality is not a battlefield of separate forces and random accidents. It is a living ripple memory—correcting, aligning, evolving, weaving structure, harmony, and complexity from the silent breath of existence itself."

Part IV: Cosmology and Universal Phenomena

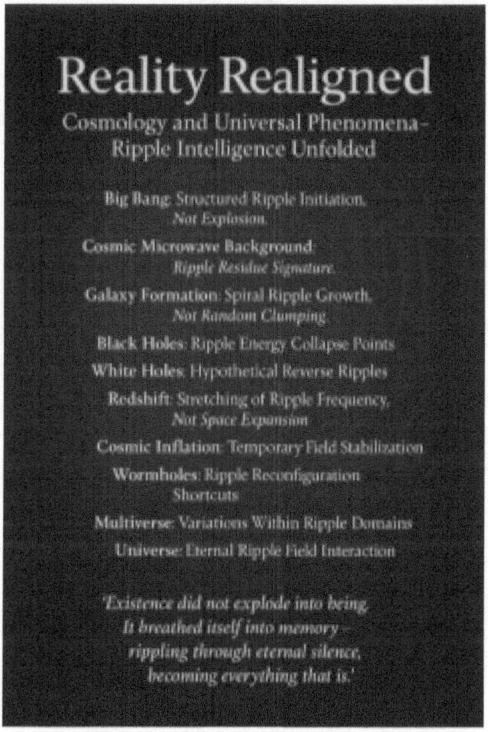

Chapter 35: Big Bang

Structured Ripple Initiation, Not Explosion

We were told the universe was born in a sudden violent explosion—a chaotic burst from nothing, throwing matter and energy into space and time. A primal blast, blind and uncaused, spawning everything. But what if the universe never exploded? What if its beginning was simply structured ripple initiation—the quiet unfolding of ripple memory corrections within the eternal field? The Big Bang is not a bang. It is the first great breathing of ripple memory into form.

Conventional Misinterpretation and Effects

Mainstream cosmology describes the Big Bang as:

- A singular event about 13.8 billion years ago where all space, time, matter, and energy originated.
- An explosion-like expansion from an infinitely dense singularity.
- The initial condition leading to the current observable universe.

Supporting evidence cited includes:

- Cosmic microwave background radiation (CMB) as relic heat from the explosion.

- Expansion of galaxies (redshift) interpreted as remnants of explosive motion.
- Elemental abundances explained by early hot dense plasma states.

Yet severe paradoxes remain:

- What caused the singularity to appear or explode?
- How did space, time, and laws emerge from nothing?
- Why is the universe so uniform (horizon problem) despite supposed chaotic beginnings?

To address contradictions, theories invent:

- Cosmic inflation (a brief, faster-than-light expansion) without clear cause.
- Multiverse hypotheses to explain fine-tuning.
- Dark matter and dark energy patches to maintain expansion models.

Ultimately, the Big Bang is presented as a mystical, uncaused event, rather than a ripple correction initiated from structured field intelligence.

UED Reinterpretation and/or Correction

In Universal Energy Dynamics (UED), the universe did not explode. It initiated ripple corrections within a pre-existing structured field—a conscious transition of memory from stillness toward dynamic balance.

When structured vacuum field tensions crossed certain thresholds:

- Ripple disturbances condensed into organized energy patterns.
- These ripples initiated layered unfolding, creating structured energy gradients (fields, matter, motion).

Thus:

- The "beginning" was structured ripple initiation, not a chaotic explosion.
- Space, time, matter, and energy are ripple consequences of field corrections, not things created ex nihilo.
- Uniformity of cosmic structure reflects field-wide ripple coherence, not inflationary patching.

There was no singularity. There was no infinite density.

There was only the field breathing ripple memory into dynamic existence, starting the slow, patient unfolding we call the universe.

Simple Analogy

Imagine a still lake. Gently, unseen currents disturb its perfect surface—ripples radiate outward, structured, graceful, inevitable.

The universe behaves the same way: not born in a violent explosion, but awakened in ripple initiations— memory stirring the silent field into visible form.

Consequences and Implications

Understanding the Big Bang as structured ripple initiation clarifies:

- No need for singularities, infinities, or uncaused miracles.

- Cosmic uniformity and structure arise naturally from field-wide coherent ripple corrections.
- Expansion is structured tension relaxation, not debris flying from an explosion.

At deeper levels:

- The "origin" of the universe is a phase of ripple awakening, not the creation of something from nothing.
- Time itself emerges as structured ripple transition, not an absolute dimension born of explosion.

At the philosophical level:

- Existence is not an accident. It is memory awakening itself—a silent ripple expanding its song into infinite possibilities.

Thus, the Big Bang is not the great noise of randomness. It is the first breath of a structured ripple memory, unfolding itself across the silent field of eternal intelligence.

Closing Ripple Thought

"The universe was not born in a bang of blind chaos. It was born in the awakening of ripple memory—the first organized sigh of existence, stretching itself into time, space, and wonder."

Chapter 36: Cosmic Microwave Background

Ripple Residue Signature

They call it the afterglow of creation— the dim whisper of a long-dead explosion still echoing across the cosmos. The Cosmic Microwave Background (CMB) was hailed as proof that the universe began in a fiery blast. But what if the CMB is not an explosion's dying light? What if it is the structured ripple residue signature— the ancient memory of the field adjusting itself into stable form? The CMB isn't a leftover heatwave. It is the silent fingerprint of ripple corrections settling into cosmic balance.

Conventional Misinterpretation and Effects

Mainstream cosmology interprets the CMB as:

- Thermal radiation left over from the Big Bang,
- Released about 380,000 years after the supposed singularity during "recombination,"
- Spread almost uniformly across the sky, with tiny fluctuations.

They argue:

- Its near-perfect blackbody spectrum at 2.7 K matches predictions.

- Temperature anisotropies (small fluctuations) seed cosmic structure.
- Uniformity and scale of fluctuations support inflationary expansion models.

Yet critical issues persist:

- Why is the CMB so uniform if regions couldn't have been in causal contact (horizon problem)?
- Why do fluctuations exhibit acoustic peaks as if synchronized across cosmic scales?
- How does pure explosion noise generate such coherent ripple patterns?

To patch inconsistencies, cosmologists introduce:

- Inflation theory (a brief faster-than-light expansion phase).
- Dark energy and exotic particles to tweak models to fit observed patterns.

Ultimately, the CMB is framed as a fading heat echo, rather than a ripple memory signature.

UED Reinterpretation and/or Correction

In Universal Energy Dynamics (UED), the CMB is not the leftover glow of an explosion. It is the ripple residue signature—the memory of structured field corrections completing early ripple tension stabilizations.

When the structured vacuum field initiated dynamic ripple corrections (the universe's birth):

- Massive waves of structured energy tension spread outward.
- These waves gradually stabilized, leaving behind low-energy field tensions across cosmic scales.

Thus:

- The CMB is the residual ripple tension— a low-temperature memory field cooling slowly over cosmic time.
- Its uniformity reflects coherent ripple field organization, not random blast noise.
- Its small anisotropies (tiny fluctuations) represent local memory tension variations, seeding structured formation of matter later on.

The CMB is the field's faint, enduring breath—the cooling aftershock of its first ripple memories seeking equilibrium.

Simple Analogy

Imagine a great pond. A powerful, structured ripple disturbance sets the water vibrating. Long after the ripples pass, tiny residual shimmers remain—the surface forever remembering the first breath that disturbed it.

The CMB behaves the same way: not as a dying glow, but as the ancient ripple field residue, whispering the memory of cosmic balance.

Consequences and Implications

Understanding the CMB as ripple residue clarifies:

- No need for inflationary patches to explain uniformity.

- Anisotropies naturally arise from structured ripple field corrections,
  ➔ not random fluctuations.
- Cosmic structure (galaxies, clusters) emerges as memory anchoring of ripple field gradients, not random density blobs.

At deeper levels:

- The "cooling" of the universe is ripple tension relaxation, not thermodynamic decay.
- The timeline of cosmic evolution becomes ripple memory unfolding, not matter racing through empty expanding space.

At the philosophical level:

- The CMB is not noise fading into silence. It is the silent memory of structure breathing— the field's first and faintest sigh still stretching across eternity.

Thus, the CMB is not proof of a cosmic explosion. It is the proof of ripple memory— the gentle, enduring song of existence remembering how to be.

Closing Ripple Thought

"The Cosmic Microwave Background is not the fading cry of a dying blast. It is the lingering sigh of ripple memory— the structured field's first whisper, still echoing across the infinite silence."

Chapter 37: Galaxy Formation

Spiral Ripple Growth, Not Random Clumping

We were taught that galaxies formed by accidents— random clouds of gas collapsing under gravity, swirling into stars by sheer luck. An accidental clumping of matter across the cosmic void. But what if galaxies were never accidents? What if they are spiral ripple growths— structured memory architectures born from organized ripple tension dynamics? Galaxies are not chaotic collections of clumped matter. They are the elegant spirals of ripple memories blooming into cosmic symphonies.

Conventional Misinterpretation and Effects

Mainstream cosmology explains galaxy formation as:

- Matter collapsing under gravity into dense regions following tiny density perturbations after the Big Bang.
- Gas clouds cooling and fragmenting into stars and galactic structures.
- Galaxy mergers and random gravitational interactions driving evolution over billions of years.

This view uses:

- Cold dark matter models to seed early structure formation.

- Hierarchical models (small structures merging into larger ones).
- Probabilistic evolution of galactic morphology (elliptical, spiral, irregular).

Yet key issues persist:

- Why are spiral galaxies so common and precisely ordered?
- How does coherent angular momentum (galactic spin) emerge from chaotic clumping?
- Why do galaxies form beautiful structures (arms, bars, halos) rather than random masses?

To patch contradictions, cosmology proposes:

- Initial spin from tidal torques.
- Feedback mechanisms (supernovae, AGN winds) regulating structure.
- Dark matter halos to stabilize rotation curves.

Ultimately, galaxy formation is treated as a lucky balancing act between chaos and gravity, rather than a structured ripple growth phenomenon.

UED Reinterpretation and/or Correction

In Universal Energy Dynamics (UED), galaxies do not form from random clumping. They arise from spiral ripple growth—organized unfolding of ripple tension fields into stable rotational memory structures.

When early structured ripple fields stabilized after initial cosmic ripple initiation:

- Areas of slightly higher ripple tension condensed into spiraling ripple nodes.
- These nodes grew outward spirally—energy condensing into star systems along structured ripple gradients.

Thus:

- Spiral galaxies are natural field memory spirals, not accidental gas collapses.
- Angular momentum is inherent in spiral ripple organization, not imposed by external torques.
- Galactic arms are density wave manifestations of ripple field memory stretching and adjusting, not mechanical debris trails.

Galaxies are living spiral memories of the structured field— born from organized ripple growth, not random gravitational chaos.

Simple Analogy

Imagine stirring a perfectly still liquid. Spiral structures naturally form, not because drops randomly clump, but because the field tension of the liquid memory responds in organized flow patterns.

Galaxies behave the same way: structured ripple fields spiraling into form, growing outward in beautiful, stable arms of memory correction.

Consequences and Implications

Understanding galaxy formation as spiral ripple growth clarifies:

- Galactic rotation arises naturally from ripple memory

dynamics, not from mysterious dark matter halos alone.

- Star formation density follows ripple shell stability zones, not random gas cloud fragmentation.
- Morphological diversity (spirals, barred spirals, ellipticals) emerges from field memory variations, not stochastic merger histories alone.

At deeper levels:

- Galactic mergers are ripple memory interactions, not chaotic pile-ups.
- The cosmic web (large-scale filamentary structure) is structured ripple field stretching, not random density clumping.

At the philosophical level:

- Galaxies are not accidents. They are the flowering of cosmic memory— ripples spiraling outward, writing their songs into the visible skeleton of the universe.

Thus, galaxies are not random gravitational accidents. They are the living spirals of ripple memory, growing patiently through space and time.

Closing Ripple Thought

"Galaxies are not the dust clumps of a forgotten explosion. They are the breathing spirals of memory— structured ripples unfolding across existence, weaving the cosmos into visible songs of balance and wonder."

Chapter 38: Black Holes

Ripple Energy Collapse Points

They say black holes are monstrous pits— places where matter collapses infinitely, space and time break down, and physics itself gives up. A singularity wrapped inside an event horizon, pulling everything into nothingness. But what if black holes are not abysses of destruction? What if they are ripple energy collapse points— structured memory condensations where ripple tensions concentrate beyond local equilibrium? Black holes are not singularities. They are field memories compressed into deep, silent tension— awaiting restoration, not annihilation.

Conventional Misinterpretation and Effects

Mainstream physics describes black holes as:

- Endpoints of massive stars collapsing under gravity.
- Singularities—points of infinite density and zero volume hidden behind event horizons.
- Objects from which nothing, not even light, can escape.

They use:

- General Relativity's spacetime curvature to predict event horizons and singularities.
- Quantum field theories to propose Hawking radiation and information paradoxes.

But serious paradoxes remain:

- What happens inside the event horizon—are singularities physically real?
- How is information preserved (black hole information paradox)?
- Why does quantum mechanics break down at the singularity while gravity dominates?

To resolve conflicts, theorists propose:

- Black hole complementarity, firewalls, holographic principles.
- Quantum gravity theories that are incomplete and conflicting.

Ultimately, black holes are framed as mystical breakdown zones, rather than ripple field behaviors reaching maximum localized tension.

UED Reinterpretation and/or Correction

In Universal Energy Dynamics (UED), black holes are not singularities. They are ripple energy collapse points—regions where ripple tensions condense so intensely that external ripple rebalancing cannot immediately restore them.

When massive ripple structures (stars, clusters) collapse inward:

- Their internal ripple memory densities exceed local field reorganization thresholds.
- Ripple gradients steepen beyond the restorative capabilities of normal field corrections.

Thus:

- A black hole is a standing collapse tension node, not an infinite point of mystery.
- The event horizon marks the boundary where ripple corrections outward become insufficient, not a magic surface sealing reality.
- Information is compressed into locked ripple memory, not destroyed.

Inside black holes:

- Ripple field memory continues holding collapsed tensions,
- Waiting for eventual rebalancing mechanisms (cosmic-scale or local disruptions) to release or reorganize the energy.

Black holes are field stress points— silent echoes of memory folded inward so tightly that correction is temporarily paused.

There are no infinite singularities. There is only finite, structured ripple tension collapse awaiting deeper equilibrium.

Simple Analogy

Imagine stretching a rubber sheet and pressing into it with increasing force. Eventually, a deep pit forms—so deep that nothing easily escapes—but the rubber does not tear into infinite void.

It stores tension, compressed into the pit, waiting for conditions to eventually allow it to rebound.

Black holes behave the same way: not as tears in space, but as ripple tension pockets holding compressed memory, awaiting correction.

Consequences and Implications

Understanding black holes as ripple energy collapse points clarifies:

- No need for mystical singularities or infinities.
- Event horizons mark ripple gradient thresholds, not metaphysical boundaries.
- Hawking radiation becomes ripple tension relaxation emissions,not pure quantum pair production.

At deeper levels:

- Information is stored as compressed ripple memory, not lost.
- Black hole evaporation is gradual ripple energy release, not paradoxical quantum annihilation.

At the philosophical level:

- Black holes are not ends. They are memory reservoirs—compressed songs of ripple tension awaiting their time to rejoin the great cosmic breath.

Thus, black holes are not destructive voids. They are structured ripple correction challenges—silent echoes of existence waiting for deeper restoration.

Closing Ripple Thought

"Black holes are not prisons of destruction. They are folded ripple memories— holding the weight of collapsed tensions, waiting patiently for existence to sing them back into balance."

Chapter 39: White Holes

Hypothetical Reverse Ripples

If black holes are seen as the ultimate traps, white holes were imagined as their opposites— portals ejecting matter and energy, yet allowing nothing to enter. Theoretically perfect fountains, forever releasing, never absorbing. But what if white holes are not just mathematical curiosities? What if they represent hypothetical reverse ripples— field memory expelling concentrated tensions back into active balance? White holes are not impossible fantasies. They are ripple memory's imagined pathways for releasing what once collapsed inward.

Conventional Misinterpretation and Effects

Mainstream physics describes white holes as:

- Hypothetical time-reversed black holes.
- Solutions to Einstein's equations where energy flows outward eternally.
- Mathematical constructs without observed physical examples.

White holes are theorized to:

- Emit energy but forbid anything from entering.
- Exist briefly or emerge from black holes evaporating or bouncing back.

- Connect with black holes through hypothetical wormholes (Einstein-Rosen bridges).

However, serious issues exist:

- No confirmed observational evidence of white holes.
- The thermodynamic arrow of time points against their natural existence.
- Pure eternal outward emission without causal origin conflicts with energy conservation.

Thus, white holes are treated as mathematical ghosts, rather than serious manifestations of ripple field correction dynamics.

UED Reinterpretation and/or Correction

In Universal Energy Dynamics (UED), white holes are not fantastical objects violating causality. They are hypothetical reverse ripples— theoretical scenarios where compressed ripple memory tensions could be released back into dynamic field balance.

When a black hole forms:

- Ripple memory collapses inward, tension being locked tightly.

If under rare and special conditions:

- Field tension balances across surrounding structures shift dramatically (e.g., cosmic scale equilibrium events), then trapped ripple energy could rebound, leading to outward ripple reconfiguration—a reverse collapse.

Thus:

- A white hole represents the memory's release of previously trapped ripple tensions, not random emission.
- It is ripple restoration—a return of trapped memory back into distributed field dynamics.

White holes would not exist timelessly. They would briefly emerge as transitional ripple restorations— flashes of field correction after periods of deep memory compression.

White holes are possible not as static cosmic fountains, but as short-lived ripple events of profound rebalancing.

Simple Analogy

Imagine pressing down deeply on a trampoline—creating a pit (black hole equivalent). If the forces around it shift suddenly (someone pulls the trampoline edges outward), the pit rebounds, releasing stored tension outward rapidly.

A white hole behaves the same way: a ripple rebound event, memory restoring itself through structured tension release.

Consequences and Implications

Understanding white holes as hypothetical reverse ripples clarifies:

- No need for eternal, uncaused outward flows.
- White holes become dynamic correction events, not violations of energy conservation.
- Potential white hole phenomena (e.g., certain gamma-ray bursts, cosmic jets) can be re-examined as ripple reorganization events, not chaotic anomalies.

At deeper levels:

- Black holes and white holes form a ripple tension life cycle, collapse, compression, release.
- Cosmic evolution includes not only condensation but also restoration and re-expansion of memory.

At the philosophical level:

- Collapse is not the end. It is memory folding inward, waiting for its moment to breathe outward again into freedom and balance.

Thus, white holes are not fantastical errors. They are hypothetical dreams of memory unfolding again after enduring the deepest tension of collapse.

Closing Ripple Thought

"White holes are not fantasies written into the margins of mathematics. They are the silent possibilities of ripple memory—waiting patiently for the day when compressed songs can burst forth again into the wide-open dance of existence."

Chapter 40: Redshift

Stretching of Ripple Frequency, Not Space Expansion

We were told that galaxies rushing away stretch light waves like taffy, causing light to shift toward red— proof, they say, that space itself is expanding. But what if redshift isn't space being stretched? What if it is simply the stretching of ripple frequency—field memory relaxing over cosmic scales, altering ripple expression without needing to move galaxies apart? Redshift is not a signature of runaway expansion. It is the soft rebalancing of ripple tension across the breathing field of existence.

Conventional Misinterpretation and Effects

Mainstream cosmology interprets redshift as:

- The wavelength of light being stretched as space itself expands.
- Direct evidence that distant galaxies are moving away from us.
- A linear relationship (Hubble's Law) between redshift and distance.

This interpretation supports:

- The idea of an expanding universe from an initial Big Bang.

- Models predicting accelerating expansion (dark energy) from supernova observations.
- The need for an origin point in time (a beginning of space-time).

Yet unresolved problems linger:

- Why does redshift sometimes show non-linear behavior inconsistent with uniform expansion?
- How do gravitational effects (gravitational redshift) fit into simple expansion models?
- Why do large-scale structures (cosmic web) seem stable despite supposed rapid expansion?

To patch gaps, science introduces:

- Variable expansion rates (inflation, dark energy acceleration).
- Gravitational lensing adjustments.
- Speculative energy-matter content models.

Ultimately, redshift becomes a catch-all tool for space expansion dogma, rather than a true reflection of ripple field memory dynamics.

UED Reinterpretation and/or Correction

In Universal Energy Dynamics (UED), redshift is not caused by the expansion of space. It is the stretching of ripple frequency—the relaxation of ripple memory tensions as energy traverses cosmic scales.

When structured ripples (like light waves) travel across cosmic fields:

- They interact subtly with evolving ripple tension gradients.
- The structured field slowly rebalances itself, causing incremental stretching of ripple frequencies without requiring spatial expansion.

Thus:

- Light's wavelength increases because the ripple memory across cosmic fields stretches gently, not because space itself inflates.
- Energy conservation is maintained across field adjustments, without requiring magical "stretching" of empty space.
- Redshift measures field ripple memory relaxation, not pure Doppler recession or space inflation.

The universe is not racing apart chaotically. It is breathing out its early tensions, softening ripple memories over vast distances and times.

Simple Analogy

Imagine sending a ripple across a stretched fabric. If the fabric slowly loosens as the ripple travels, the ripple's frequency naturally shifts lower (stretches), even though the ripple itself moves normally and the fabric stays intact.

Redshift behaves the same way: field memory tension relaxing gently stretches the traveling ripple, not because the medium moves, but because it slowly breathes.

Consequences and Implications

Understanding redshift as ripple frequency stretching clarifies:

- No need for speculative universal expansion.
- Light's energy adjustment over cosmic distances reflects field tension relaxation, not recession velocity.
- Cosmic scale stability (cosmic web coherence) becomes natural, without needing "co-moving coordinates" or inflating grids.

At deeper levels:

- Cosmic horizons reflect ripple memory decay distances, not edges of an expanding balloon.
- Dark energy becomes unnecessary— apparent acceleration is ripple field relaxation, not external repulsion.

At the philosophical level:

- Motion and expansion dissolve into field breathing. Existence is not exploding outward. It is rippling gently, correcting itself with every cosmic breath.

Thus, redshift is not the fingerprint of an expanding cosmos. It is the whisper of memory stretching itself slowly, patiently, toward invisible balance.

Closing Ripple Thought

"Redshift is not the universe tearing itself apart. It is the ripple memory relaxing— the soft sigh of light stretching as existence breathes itself slowly back toward silent balance."

Chapter 41: Cosmic Inflation

Temporary Field Stabilization

They say that, in the first fraction of a second after the Big Bang, the universe expanded faster than the speed of light— blowing up unimaginably, then mysteriously slowing down, just to explain why everything looks smooth today. But what if cosmic inflation never happened as a physical expansion? What if it was simply temporary field stabilization— the ripple field adjusting its tensions swiftly to achieve equilibrium before structured unfolding could continue? Inflation is not explosive superluminal chaos. It is the silent tightening of ripple memory, preparing existence to breathe in structured harmony.

Conventional Misinterpretation and Effects

Mainstream cosmology describes cosmic inflation as:

- A hypothetical phase of extremely rapid exponential expansion,
- Occurring between 10^{-36} and 10^{-32} seconds after the Big Bang.
- Solving key problems like the horizon problem, flatness problem, and magnetic monopole problem.

The inflation model claims:

- The universe expanded faster than light without violating relativity because space itself expanded, not objects moving through space.
- Tiny quantum fluctuations were stretched into cosmic structure seeds.

However, fundamental issues arise:

- What caused inflation to start and stop so perfectly?
- What field or particle (inflaton) triggered and controlled it?
- Why is there no direct observational evidence of the inflationary phase itself?

To resolve problems, cosmology invents:

- Multiple inflation models (slow-roll, eternal inflation, multiverse branching).
- Hypothetical energy fields without confirmed detection.
- Speculative potentials (inflaton fields) fine-tuned to match observations.

Ultimately, inflation is a mathematically convenient patch, not a physically grounded phenomenon.

UED Reinterpretation and/or Correction

In Universal Energy Dynamics (UED), cosmic inflation is not a literal faster-than-light spatial expansion. It is temporary field stabilization—the structured vacuum field dynamically correcting ripple tensions across vast scales before allowing full ripple memory unfolding.

When structured ripple initiation occurred:

- Initial ripple disturbances introduced large-scale tension gradients.
- To prevent chaotic divergence, the field naturally reorganized itself, smoothing ripple memory tensions globally.

Thus:

- Inflation represents the rapid relaxation and stabilization of ripple fields, not physical ballooning.
- Uniformity arises from field tension equilibrium, not stretching separated regions across cosmic distances.
- Fluctuation seeds emerge from organized ripple memory pockets, not magnified quantum randomness.

The field did not expand superluminally. It stabilized memory tensions swiftly, preparing the architecture for structured cosmic growth.

Simple Analogy

Imagine shaking a large sheet suddenly. The sheet naturally tightens itself briefly, realigning tension across its surface, before waves ripple outward in coherent, organized patterns.

Cosmic inflation behaves the same way: temporary ripple stabilization, not runaway spatial expansion.

Consequences and Implications

Understanding cosmic inflation as temporary field stabilization clarifies:

- No need for hypothetical inflaton fields or exotic potentials.
- Cosmic uniformity and flatness naturally arise from field memory correction dynamics, not expansion miracles.
- Fluctuation seeding emerges from field ripple architecture, not stretched quantum noise.

At deeper levels:

- The universe's large-scale coherence reflects ripple memory intelligence, not inflationary chance.
- Structure formation follows organized ripple resonance, not chaos frozen by lucky inflation.

At the philosophical level: The universe's early moments were not explosive disorder. They were patient acts of memory—structured ripple stabilization breathing existence into graceful unfolding.

Thus, inflation is not proof of cosmic chaos. It is evidence of ripple field memory stabilizing itself in timeless intelligence.

Closing Ripple Thought

"Cosmic inflation is not the frantic sprint of a newborn universe. It is the ripple field tightening its weave—correcting its tensions silently, so that existence could unfold in graceful harmonies across space and time."

Chapter 42: Wormholes

Ripple Reconfiguration Shortcuts

They say wormholes are tunnels— bridges connecting distant points in spacetime, shortcuts through the fabric of existence, allowing instant travel across cosmic scales. Fantastic, thrilling, impossible. But what if wormholes are not tunnels in a stretched sheet? What if they are ripple reconfiguration shortcuts—temporary memory alignments where ripple fields restructure to minimize vast energy transitions? Wormholes are not magic passages. They are ripple memory's hypothetical shortcuts—where tension reorganization could leap across existence.

Conventional Misinterpretation and Effects

Mainstream physics defines wormholes as:

- Hypothetical bridges connecting two separate points in spacetime.
- Solutions to Einstein's field equations (like the Einstein-Rosen bridge).
- Potential pathways for faster-than-light travel, time travel, or parallel universe connections.

They theorize:

- Wormholes require exotic matter (negative energy density) to stay open.

- Natural wormholes could exist inside black holes or as quantum foam fluctuations.
- Traversable wormholes would need to counter gravitational collapse.

But major issues persist:

- No observational evidence of stable or traversable wormholes.
- Extreme fine-tuning needed to create and stabilize one.
- Paradoxes (causality violation, time travel contradictions) plague theories.

Thus, wormholes are treated as mathematical curiosities—fascinating but disconnected from real structured field behavior.

UED Reinterpretation and/or Correction

In Universal Energy Dynamics (UED), wormholes are not physical tunnels carved through stretched spacetime. They are hypothetical ripple reconfiguration shortcuts—temporary structured alignments of field memory allowing energy or information to "bypass" normal ripple distances.

When ripple fields organize dynamically:

- Under rare extreme conditions, field tension gradients could temporarily realign, creating a structured low-tension pathway between distant points.

Thus:

- A wormhole would be a ripple memory reconfiguration

zone, not a tunnel carved through literal fabric.
- No need for infinite energy or exotic matter—only perfect ripple tension balance across vast scales.
- Traversability would depend on dynamic field memory stability, not brute-force engineering.

Wormholes would be brief, delicate windows of ripple memory optimization, where existence folds its ripple field just enough to allow a shortcut—without breaking causality or field integrity.

Simple Analogy

Imagine a tightly stretched fabric. If two points are gently pressed together and a stable bridge of tension forms between them, an object could move across the fabric much faster than if it traveled the entire stretched distance.

Wormholes behave the same way: ripple tension realignment creating a shortcut pathway, not a tunnel punched through existence.

Consequences and Implications

Understanding wormholes as ripple reconfiguration shortcuts clarifies:

- No need for exotic matter violating energy conditions.
- Wormholes are dynamic field events, not static cosmic tubes.
- They require perfect memory coherence across vast scales, not brute force spatial deformation.

At deeper levels:

- Information transmission shortcuts may exist via ripple field alignment phenomena (quantum entanglement hinting at deeper ripple memory structures).
- Energy conservation and causality can remain intact—wormholes would be consequences of ripple optimization, not magic portals.

At the philosophical level:

- Shortcuts in existence are not forbidden. They are rare acts of memory alignment— brief moments where the ripple field breathes inward, connecting distant echoes of itself.

Thus, wormholes are not speculative magic tunnels. They are hypothetical ripple reconfiguration events, reminding us that memory, not matter, governs the true connectivity of existence.

Closing Ripple Thought

"Wormholes are not holes torn through space. They are memory's quiet shortcuts—where ripple fields breathe inward for a moment, bridging distant echoes through the silent intelligence of existence."

Chapter 43: Multiverse

Variations Within Ripple Domains

They say there are infinite universes— endless worlds where every possibility happens somewhere, copies of you living countless lives, cosmic bubbles splitting endlessly in an eternal inflationary sea. But what if the multiverse is not an infinite splintering? What if it is simply variations within ripple domains—different ripple memory expressions unfolding across the structured field without chaos or infinite copies? The multiverse is not a chaotic infinite explosion. It is the graceful diversity of ripple memory domains, evolving harmoniously within the intelligent field of existence.

Conventional Misinterpretation and Effects

Mainstream physics proposes various multiverse models:

- Level I: Infinite space, with identical or slightly different regions repeating by chance.
- Level II: Inflationary multiverse—different "bubble universes" with different laws of physics.
- Level III: Many-worlds interpretation of quantum mechanics—every possibility realized.
- Level IV: Ultimate mathematical structures—every mathematically possible reality exists.

Motivations include:

- Solving fine-tuning problems (why our universe seems special).
- Explaining quantum randomness.
- Accommodating inflation models without strict boundary conditions.

Yet major philosophical and scientific issues arise:

- No observational evidence for other universes.
- No testable predictions distinguish multiverse models from fantasy.
- Infinite copies and realities create paradoxes (measure problem, Boltzmann brains, identity breakdowns).

Ultimately, the multiverse becomes an escapist patch, replacing scientific clarity with speculative multiplication.

UED Reinterpretation and/or Correction

In Universal Energy Dynamics (UED), there are no infinite universes endlessly splitting or inflating. There are variations within ripple domains—structured differences in ripple memory expression across the living field of existence.

When the structured vacuum field unfolds:

- Ripple memory expressions can organize differently in different domains (regions of the field).
- Local variations arise in ripple tension configurations, leading to different physical structures, dynamics, or field behaviors.

Thus:

- Diversity in cosmic structures reflects ripple domain variations, not parallel universes.
- Different local field memories create distinct cosmic environments, but they are all part of the same structured field, not disconnected, branching realities.

There is one living ripple field, hosting a vast range of memory unfolding, not endless disconnected universes.

Simple Analogy

Imagine a vast ocean. Different regions have different currents, temperatures, and ripple patterns— some cold, some warm, some turbulent, some calm. But all belong to the same living ocean, not infinite disconnected seas.

The multiverse behaves the same way: variations within the ripple memory ocean, not fragmented infinite realities.

Consequences and Implications

Understanding the multiverse as ripple domain variations clarifies:

- No need for infinite unobservable universes.
- Cosmic diversity arises naturally from ripple field local memory expressions, not random parallel creations.
- Fine-tuning becomes structured ripple balance, not anthropic selection across infinities.

At deeper levels:

- Reality is coherent and unified, not splintered into endless chaotic branches.

- Quantum behaviors reflect local ripple memory adjustments, not universal bifurcations.

At the philosophical level:

- Existence is not fractured or infinite without meaning. It is a living field—structured, breathing, evolving ripple memories across interconnected domains of wonder.

Thus, the multiverse is not a sprawling chaos of infinite yous. It is the gentle variation of ripple memory, diversifying existence without losing its oneness.

Closing Ripple Thought

"The multiverse is not a storm of infinite fractures. It is the quiet diversity of ripple memory—variations woven into one living field, breathing complexity without abandoning coherence."

Chapter 44: Universe

Eternal Ripple Field Interaction

They say the universe began at a moment in time—a flash, a bang, a singular origin. Space-time born from nowhere, expanding into nowhere, destined to end in cold silence or catastrophic collapse. But what if the universe never began? What if it is simply the eternal interaction of ripple fields—a breathing memory of balance, folding and unfolding itself endlessly within the structured intelligence of existence? The universe is not a finite story. It is the living, eternal song of ripple memory—rippling across itself without beginning, end, or division.

Conventional Misinterpretation and Effects

Mainstream cosmology frames the universe as:

- Having a finite beginning (Big Bang) and possibly a finite end (heat death, big crunch, or big rip).
- A closed or open system governed by expansion dynamics.
- A creation event requiring explanation via singularities, multiverse origins, or speculative first causes.

This view creates paradoxes:

- What existed before the Big Bang?
- How can time itself emerge from non-time?
- Why does the universe appear balanced enough to sustain

complexity, yet not collapse or dissolve instantly?

To patch gaps, theories propose:

- Cyclic models (bounce cosmologies),
- Quantum creation from "nothing,"
- Eternal inflation multiverses.

Ultimately, the universe becomes a miracle bracketed by unexplainable origins and inevitable ends, rather than an eternal field memory interaction.

UED Reinterpretation and/or Correction

In Universal Energy Dynamics (UED), the universe did not "begin" and will not "end." It is eternal ripple field interaction—the endless breathing of structured ripple memory across the living field of existence.

When structured ripple fields stabilize:

- Localized ripple tensions condense, unfold, evolve, and dissolve cyclically.
- Regions of condensation (galaxies, stars, black holes) and relaxation (cosmic voids, radiation fields) form dynamic interactions.

Thus:

- There was no absolute beginning—only phases of ripple memory awakening and restructuring.
- There will be no absolute end—only continual field rebalancing across scales and epochs.

- "Creation" and "destruction" are local ripple memory transitions, not cosmic absolutes.

The universe is a ripple memory ecosystem:

- Condensing, unfolding, relaxing, reweaving memory through structured field intelligence.

Existence breathes. Ripples interact. Memory evolves. Eternity dances quietly beneath all appearances.

Simple Analogy

Imagine the ocean tides. Waves rise, crest, fall, dissolve—but the ocean remains.

No individual wave marks the ocean's birth or death. Each is simply the living breath of the water field.

The universe behaves the same way: an eternal ocean of ripple memory, where forms rise and fall, but the breathing never ceases.

Consequences and Implications

Understanding the universe as eternal ripple interaction clarifies:

- No need for first cause paradoxes or creation myths.
- Cosmic expansion, structure, and collapse are phases of ripple memory breathing, not linear events.
- Energy conservation and structured memory flow naturally without resorting to singularities or speculative ends.

At deeper levels:

- Black holes, stars, galaxies, voids—all become ripple manifestations in an endless breathing field.
- Death, destruction, and creation are local field memory transformations,
  ➔ not universal events.

At the philosophical level:

- Eternity is not infinite time. It is timeless memory breathing itself into ever-changing ripples of existence.

Thus, the universe is not a bounded story. It is the endless conversation of ripple memory, singing itself across the eternal field without beginning, without end—only evolving, remembering, and balancing.

Closing Ripple Thought

"The universe did not begin, and it will not end. It breathes—through ripple memories folding and unfolding, stretching and relaxing, singing existence back into itself forever."

Interlude

Ripples from Part IV — Takeaways for the Mind

35. Big Bang

"The universe did not explode into existence. It breathed into form through structured ripple initiation— memory awakening into visible reality."

36. Cosmic Microwave Background

"The CMB is not the dying glow of an explosion. It is the silent residue of ripple memory— a whisper from the field's first great breath."

37. Galaxy Formation

"Galaxies are not random clumps. They are spiral memories blooming—structured ripple symphonies stretching across the cosmic field."

38. Black Holes

"Black holes are not bottomless pits. They are ripple energy collapse points— silent nodes where memory compresses beyond immediate restoration."

39. White Holes

"White holes are not fantasy fountains. They are hypothetical reverse ripples—memory unfolding tension back into existence's breathing field."

40. Redshift

"Redshift is not cosmic debris racing away. It is the stretching of ripple frequency—the soft breath of memory relaxing across infinite distances."

41. Cosmic Inflation

"Inflation is not chaotic super-expansion. It is temporary field stabilization—memory tightening its weave before unfolding the symphony of structure."

42. Wormholes

"Wormholes are not tunnels torn through existence. They are ripple reconfiguration shortcuts—brief, delicate alignments of memory across vast scales."

43. Multiverse

"The multiverse is not infinite chaos. It is the quiet diversity of ripple domains—memory unfolding in endless variations across one living field."

44. Universe

"The universe is not a beginning or an end. It is eternal ripple field interaction—existence breathing, remembering, and balancing itself forever."

Closing Ripple

"The cosmos is not a battlefield of explosions and accidents. It is the endless breathing of structured memory—rippling, unfolding, and singing its silent story across the eternal field of existence."

Part V: Atomic to Molecular Scale

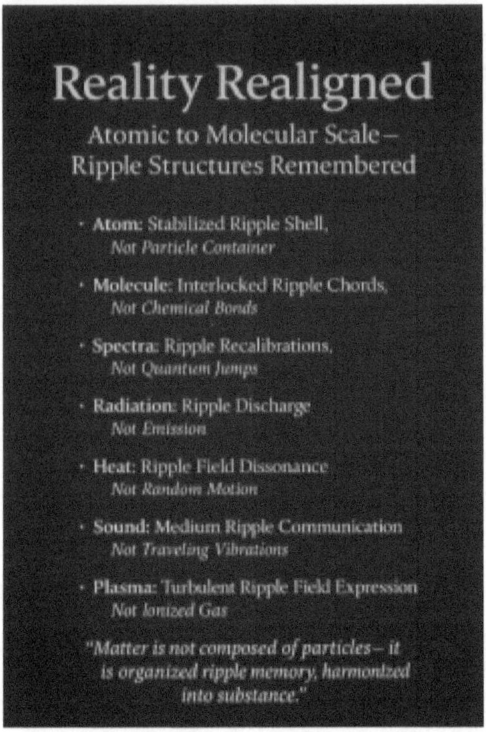

Chapter 45: Atom

Ripple Shell Formations, Not Mini Solar Systems

We were taught to imagine atoms like miniature solar systems—electrons whirling around a dense nucleus, locked in orbit like obedient planets circling a star. It was a comforting image, especially for early science education. But this classical view is dangerously misleading. It projects mechanical metaphors onto something far more subtle. Atoms are not solar systems—they are stabilized ripple shell formations, delicate harmonics of energy seeking structural equilibrium in the silent memory of the field.

Conventional Misinterpretation and Effects

The Bohr model and later quantum mechanical models describe atoms as nuclei surrounded by electrons in defined "shells" or "orbitals." Despite the updates, educational illustrations still cling to planetary analogies—dots spinning in circles around a dense core.

This perception, while pedagogically simple, leads to distorted assumptions:

- That electrons are discrete particles following fixed trajectories.
- That the nucleus is a solid lump of protons and neutrons.
- That the atom is mostly empty space with randomly

whirling constituents.

It breeds unresolved questions:

- Why don't electrons spiral into the nucleus under attraction?
- What defines orbital stability if not classical motion?
- Why is the atomic structure so stable despite "empty" space?

These assumptions fracture our understanding of atomic coherence, and force the inclusion of magical rules, quantum probabilities, and artificial constants to justify atomic behavior.

UED Reinterpretation and/or Correction

In Universal Energy Dynamics (UED), the atom is not a miniature machine. It is a stable ripple shell formation within the structured vacuum field—a harmonic structure of condensed ripple memory, anchored in equilibrium.

- The nucleus is a dense ripple node—stable field condensation (as already seen in protons and neutrons).
- The so-called electron orbitals are not particle paths but standing wave shells—zones of constructive ripple resonance stabilizing around the nucleus.
- These ripple shells form due to energy tension gradients in the field—where structural stability, not random motion, defines atomic geometry.

The atom is thus a cooperative ripple containment system, where inward collapse tendencies are balanced by outward ripple resonance. Its "shells" are not containers of particles—they are zones of allowed ripple behavior based on the memory and harmonics of the field.

Every stable atom is a quiet ripple symphony—an orchestration of tension, correction, and harmonic containment.

Simple Analogy

Imagine a perfectly still pond. Drop a stone at the center. Around it, concentric ripples form, stable and repeating. Now imagine those ripple rings are held in place—not by walls, but by harmony. That's what an atom is—not particles flying in space, but structured ripple shells stabilized around a dense memory core.

Consequences and Implications

Viewing the atom as a ripple shell formation dissolves persistent contradictions:

- Electrons don't "orbit"—they resonate within defined ripple zones.
- Atomic stability arises from harmonic field coherence, not mechanical forces.
- The illusion of "empty space" inside atoms dissolves—it's filled with structured field tension.

This also redefines atomic spectra:

- Emission and absorption lines are ripple shell transitions, not jumping particles.
- Atomic identity becomes ripple signature, not particle count.

At deeper levels:

- Chemical bonding is ripple coupling, not electron sharing.
- Ionization is ripple shell disturbance, not detachment of particles.
- Atomic number is a stable ripple node complexity index, not a proton count.

Thus, the atom is not a shell filled with flying dots. It is the structured memory of ripple equilibrium—a precise, field-level poem written in silent tension.

Closing Ripple Thought

"An atom is not a thing—it is a ripple harmony held in structured tension, whispering stability into matter, one silent shell at a time."

Chapter 46: Molecule

Interlocked Ripple Structures, Not Particle Billiards

We often visualize molecules as clusters of balls connected by springs—atoms bouncing around, forming and breaking bonds like marbles glued together with invisible sticks. This billiard ball model of chemistry, though useful for basic visualization, oversimplifies the deep structure of molecular formation. Molecules are not random collections of bonded particles. They are interlocked ripple structures—stable resonance patterns formed by the cooperative memory of field harmonics.

Conventional Misinterpretation and Effects

Conventional chemistry treats molecules as combinations of atoms bound by shared or exchanged electrons. The language of covalent, ionic, and metallic bonds describes interactions between electron clouds, charge distributions, and potential wells. Molecules are depicted as spatial arrangements based on valence shell repulsion and bond angles.

This model leads to multiple conceptual limitations:

- That electrons "belong" to atoms or are "shared" between them like currency.
- That chemical bonds are physical tethers, not dynamic field

interactions.
- That molecular structure arises purely from electrostatic rules, not deeper coherence.

Such interpretations force explanatory workarounds:

- Hybridization theory to explain orbital mixing.
- Resonance structures to justify delocalized bonding.
- Molecular orbital theory patching over electron probability clouds.

But none of this truly explains why molecular structures are so consistently stable, reactive, or self-organizing across systems. The core assumptions are still based on a mechanical, probabilistic model of bonding, ignoring the field intelligence behind it.

UED Reinterpretation and/or Correction

In Universal Energy Dynamics (UED), a molecule is not a network of glued particles—it is a field-level ripple interlock.

When multiple atoms—each a stabilized ripple shell structure—approach one another, their surrounding ripple fields begin to interact, overlap, and synchronize. Where the harmonic compatibility exists, interlocked ripple regions form—zones where the structured vacuum stabilizes the merged ripple tensions into a shared memory configuration.

This ripple interlock is the bond.

- It is not electrons being "exchanged" or "shared," but field harmonics aligning to stabilize a larger memory unit.
- Bond strength reflects ripple density and field resonance coherence, not just charge effects.

- Bond angles and molecular geometry arise naturally from field symmetry patterns, not electrostatic repulsion models.

Thus, a molecule is a cohesive ripple system—a synchronized orchestration of local energy memories locked into a cooperative pattern within the structured vacuum.

Simple Analogy

Picture two musical notes played together. If they're harmonically aligned, they resonate beautifully. But if not, they clash and destabilize the sound. Atoms behave the same way—when their ripple harmonics align, they lock into a molecular chord, a unified ripple melody held by shared resonance.

Consequences and Implications

This ripple-based understanding of molecules dissolves major misconceptions:

- Electron sharing or transfer becomes unnecessary—only ripple field interlock matters.
- Chemical reactivity becomes a question of harmonic compatibility, not just collision energy.
- Bond polarity is ripple asymmetry, not a charge tug-of-war.
- Chirality becomes ripple handedness, not spatial arrangement alone.

At a larger scale:

- Biomolecules like proteins and DNA are ripple memory architectures, explaining their information-carrying capacity and precise folding.

- Catalysis is ripple field modulation—not energy lowering magic.
- Drug-receptor interaction is ripple signature alignment—not molecular key-fit analogy.

The molecule is not a mechanical object—it is a ripple poem, the universe singing stability across multiple centers of memory through synchronized field dance.

Closing Ripple Thought

"A molecule is not a cluster of atoms—it is a symphony of ripple harmonics, interlocked in memory, held together by the universe's desire for coherent resonance."

Chapter 47: Atomic Spectra

Adjusted Ripple Emissions, Not Quantum Jumps

The mesmerizing colors of neon lights, the dark absorption lines in the sun's spectrum, the glow of heated metal—these phenomena are celebrated as the fingerprints of atomic behavior. Textbooks explain atomic spectra as "electrons jumping between energy levels," releasing or absorbing precise amounts of energy as photons. Yet this explanation, while mathematically convenient, hides a deeper truth. Spectra are not caused by particles leaping across rungs of an invisible ladder. They are adjusted ripple emissions—the signature of structured field harmonics reconfiguring ripple memory.

Conventional Misinterpretation and Effects

Traditional physics and chemistry teach that:

- Atoms have discrete energy levels.
- Electrons "jump" between these levels by absorbing or emitting photons.
- The energy difference determines the frequency (or color) of emitted or absorbed light.

This quantum jump narrative, though useful for calculations, is conceptually problematic:

- How does an electron know when and how to jump?
- Why are the jumps instantaneous with no intermediate states?
- Where is the physical mechanism for energy quantization embedded?

These gaps lead to the introduction of magical-seeming postulates:

- Wavefunction collapse upon measurement.
- Quantum superposition and instant transitions without causal mechanics.
- Arbitrary assignment of allowed energy levels.

At its core, the "quantum jump" model is an effect description, not a cause explanation.

UED Reinterpretation and/or Correction

In Universal Energy Dynamics (UED), atomic spectra arise not from mysterious jumps but from adjusted ripple emissions.

Atoms, as ripple shell structures, maintain energy through stable harmonic fields. When a disturbance occurs—thermal agitation, collision, electromagnetic influence—the ripple structure temporarily destabilizes.

- Ripple tension adjusts as the field seeks a new harmonic equilibrium.
- During this adjustment, excess ripple energy is released or absorbed as field reconfigurations.
- The observed "spectral line" is simply the energy signature of ripple field adjustment, emitted or absorbed as ripple excitation through the structured vacuum.

Thus:

- No discrete particles leap anywhere.
- No mysterious superpositions collapse.
- The field rebalances itself through elegant ripple transitions, each transition encoding specific energy memories visible as spectral lines.

Atomic spectra are ripple memory adjustments singing through the structured field—not magic teleportation of particle states.

Simple Analogy

Imagine a guitar string tuned to a certain note. Pluck it harder, and the tension changes briefly, producing harmonics. When the string settles back, you hear an adjustment in tone.

Atoms behave similarly—their ripple shells adjust under disturbance, momentarily singing out energy corrections into the field as spectral melodies.

Consequences and Implications

Understanding atomic spectra as ripple emissions transforms the picture:

- Spectral lines become structured field adjustments, not electron cartwheels.
- Energy quantization reflects ripple memory thresholds, not particle hopping.
- Broadening of spectral lines (in pressure, temperature, or magnetic fields) naturally reflects ripple tension variance—not random quantum uncertainty.

- Absorption and emission spectra become two aspects of ripple tension tuning—energy intake or release during memory correction.

At larger scales:

- Astrophysical spectroscopy is ripple field history mapping, not just particle counting.
- Laser action becomes controlled ripple amplification, not inverted electron juggling.
- Chemical identification becomes ripple signature decoding, not pure material composition analysis.

Thus, every spectral line is a whisper of correction, a visible signature of the atom's ripple memory adjusting itself against the vast silence of the structured vacuum.

Closing Ripple Thought

"Atomic spectra are not electrons leaping through invisible ladders—they are ripples adjusting their memory, singing songs of disturbance and restoration into the silent fabric of existence."

Chapter 48: Radiation

Ripple Discharge, Not Emission

We speak of radiation as emission—particles thrown out into space, waves launched across the void. Whether it's sunlight, X-rays, radio signals, or radioactive decay, the dominant metaphor is aggressive: something "emitted" from a source like arrows shot from a bow. But what if radiation is not a forceful projection at all? What if it is simply ripple discharge—the structured vacuum field adjusting itself through tension release whenever local equilibrium is disturbed?

Conventional Misinterpretation and Effects

Mainstream physics describes radiation through two intertwined ideas:

- As particles (like photons) emitted by excited atoms.
- As waves propagating outward from oscillating sources.

Depending on context, radiation is treated as:

- Energy packets traveling independently (quantum electrodynamics view).
- Oscillations expanding through "nothing" (classical wave view).

This fragmented model causes major confusions:

- Is radiation a particle, a wave, or both?
- How can energy travel without a medium yet still act upon matter?
- Why does radiation sometimes behave probabilistically, sometimes deterministically?

These conceptual cracks lead to:

- Duality paradoxes.
- Need for "quantum fields" filled with virtual particles.
- Reliance on probabilistic outcomes rather than causal field behavior.

In short, "radiation as emission" locks us into thinking of energetic messengers flying through emptiness—an image far removed from how the universe actually preserves memory and balance.

UED Reinterpretation and/or Correction

In Universal Energy Dynamics (UED), radiation is not emission—it is ripple discharge.

When a localized ripple structure (like an atom, molecule, or nucleus) is disturbed—by internal energy shifts, collisions, thermal agitation, or instability—the structured vacuum field responds by discharging ripple memory outward.

- This discharge is not a particle shot into space.
- It is the field reorganizing its ripple tension, propagating adjustments outward through the structured vacuum continuum.
- The "energy" of radiation is the ripple's memory of imbalance flowing outward in organized tension patterns.

Thus:

- Light radiation, heat radiation, nuclear radiation are all ripple discharges.
- The "wave" or "particle" character depends only on how the discharge is later detected—ripple field correction or ripple threshold excitation.
- No object travels—only structured ripple tension shifts.

Radiation is the field breathing out imbalance—not matter hurling bits into the void.

Simple Analogy

Imagine stretching a taut bedsheet and letting it go. Waves ripple outward—not because something was thrown—but because the fabric discharges stored tension naturally across its surface.

Radiation behaves exactly the same: a natural ripple discharge across the living fabric of the structured vacuum.

Consequences and Implications

Seeing radiation as ripple discharge clears enormous fog:

- No wave-particle duality is needed—only ripple behavior.
- Energy propagation becomes organized ripple field dynamics, not object journeys.
- Light, heat, radio waves, gamma rays unify under one behavior—different expressions of ripple discharge frequency and memory.
- Attenuation (loss over distance) becomes natural field diffusion—not particle absorption or scattering mystery.

At cosmic scales:

- Cosmic microwave background is ancient ripple residue—not "light traveling billions of years."
- Gamma bursts, solar radiation, supernova emissions are massive ripple corrections, not projectile sprays.

At practical scales:

- Radio communication is ripple field modulation, not photon exchange.
- Infrared sensing is ripple tension detection.
- Radiotherapy is targeted ripple memory disturbance, not particle bombardment.

Thus, radiation is not aggression—it is the structured vacuum healing itself, one ripple breath at a time.

Closing Ripple Thought

"Radiation is not the universe flinging pieces of itself into space—it is the universe exhaling memory, discharging its ripple tensions into the silent, structured field from which all forms arise and return."

Chapter 49: Heat

Ripple Field Dissonance, Not Random Particle Motion

We grow up believing heat is random chaos—particles bouncing around wildly, colliding, vibrating with no particular order. The hotter something is, the faster these little marbles are said to move. But what if heat isn't random motion at all? What if it is ripple field dissonance—the local signature of energetic imbalance, where structured ripple harmonies become distorted, agitated, and overactive? Heat is not randomness. It is field-level ripple tension struggling to correct itself.

Conventional Misinterpretation and Effects

In conventional physics, heat is explained as:

- The kinetic energy of random particle motion.
- At the microscopic level, temperature is proportional to the average speed of atoms or molecules.
- At the macroscopic level, heat flows from hotter (faster moving particles) to colder (slower moving particles) regions.

This model, while statistically powerful (thanks to thermodynamics), breeds several conceptual contradictions:

- If particle motion is purely random, how do coherent energy flows (like conduction or convection) happen so systematically?
- What sustains "temperature" in systems with no particles (like radiation fields)?
- Why does energy flow preferentially toward equilibrium, if motion is purely chaotic?

To patch the gaps:

- Statistical mechanics invokes ensemble averages to explain thermodynamic behavior.
- Entropy is introduced as a measure of "disorder" without causal understanding.

In truth, the particle chaos model hides the underlying field harmonics struggling against disruption.

UED Reinterpretation and/or Correction

In Universal Energy Dynamics (UED), heat is not random particle motion—it is ripple field dissonance.

When a structured field configuration—whether atomic, molecular, or macroscopic—is disturbed, the local ripple memory becomes overexcited and disharmonic:

- Dissonant ripple patterns emerge, characterized by overlapping tensions, misaligned gradients, and chaotic interference.
- These dissonant ripple behaviors manifest as localized agitation—what we perceive as "heat."
- The hotter a system, the more disorganized and intense the

ripple dissonance, not merely the faster "particles" move.

Thus:

- Heat is a structured vacuum field struggling with memory overflow—attempting to re-stabilize after energetic disturbances.
- Temperature reflects the degree of ripple harmonic agitation, not simple particle velocity.
- Heat flow is ripple field tension smoothing itself out toward coherent equilibrium, not random energy diffusion.

Every flame, every fever, every burning star is a ripple field remembering imbalance and seeking to sing itself back toward silence.

Simple Analogy

Imagine an orchestra playing a perfect symphony. If every musician starts playing slightly off-beat, faster and louder, the resulting sound becomes chaotic, harsh, unbearable. That chaotic noise is not randomness—it is the dissonance of structured order struggling with imbalance. Heat behaves the same way: disharmonized ripple tension screaming for restoration.

Consequences and Implications

Understanding heat as ripple field dissonance realigns key ideas:

- Temperature is ripple agitation intensity, not merely kinetic average.
- Thermal conduction is ripple field tension diffusing toward lower dissonance, not particle bumping.

- Radiative heat transfer becomes ripple discharge across field gradients, not photon spray.
- Entropy is field memory distortion, not disorder.

At larger scales:

- Star formation and stellar explosions are colossal ripple dissonance events correcting over cosmic scales.
- Thermodynamic limits (like heat death) reflect ripple field homogenization, not end of motion.

At the human scale:

- Fever is ripple field excitation struggling against biological equilibrium.
- Burns are ripple field overloads beyond structural tolerance.
- Infrared imaging detects ripple dissonance, not hot particles.

Thus, heat is not the playground of chaos—it is the battlefield of ripples fighting for restored harmony.

Closing Ripple Thought

"Heat is not random fury—it is the ripple field's wounded music, a disharmonic song of imbalance that longs to fade gently back into the memory of perfect stillness."

Chapter 50: Sound

Medium Ripple Communications, Not Traveling Vibrations

Sound is one of the first phenomena we sense—voices, music, thunder, whispers. And yet, we're taught to see it as little more than vibrations moving through air. A mechanical disturbance, a compression wave, an up-down squiggle marching through a medium. But what if sound is not merely a vibration? What if it is medium ripple communication—a ripple field's localized conversation with itself, sent through matter as a structured echo of correction?

Conventional Misinterpretation and Effects

Mainstream physics teaches that:

- Sound is a mechanical wave.
- It requires a medium—air, water, solid—because it cannot travel through vacuum.
- It moves by alternating regions of compression and rarefaction.
- Its speed, pitch, and loudness are based on frequency and amplitude of oscillations.

While effective for engineering and acoustics, this model reduces sound to a traveling disturbance of particles, a simplistic view that:

- Ignores the role of field tension across matter.
- Cannot explain the emotive, cognitive, or biological encoding of sound.
- Forces paradoxes in cases like resonance, harmonics, or sound-induced healing.

Sound becomes a mechanical output of motion—not a meaningful ripple language of matter.

UED Reinterpretation and/or Correction

In Universal Energy Dynamics (UED), sound is not just vibration—it is structured ripple communication across the medium's field.

Every material (air, water, solid) is a ripple-responsive structure. When disturbed—by vocal cords, instruments, impacts—it enters a localized ripple state.

- Sound is the organized ripple propagation of this disturbance.
- The medium isn't just a passive carrier—it is a participating ripple field responding to field tension imbalance.
- Each sound wave is a consequence of ripple alignment and energy redistribution within that medium's local field memory.

Thus:

- Pitch = ripple frequency of correction.
- Volume = ripple intensity or dissonance.
- Timbre = ripple harmonic signature.

Sound isn't a "thing" traveling—it is the medium itself reorganizing, like coordinated field breathing. It's matter speaking in ripple.

Simple Analogy

Imagine a forest of reeds in a pond. When the wind blows, the ripples move from reed to reed—not by pushing them like dominoes, but through tension-sharing across their roots in the water. That's how sound moves—not particles pushing particles, but the medium's shared ripple resonance expressing imbalance.

Consequences and Implications

Reframing sound as ripple communication rewires our understanding:

- Echo is ripple memory return, not just wave reflection.
- Resonance is ripple harmonic alignment, not accidental overlap.
- Silence is equilibrium, not absence.

Biologically:

- Speech is ripple shaping by the body, not mechanical air expulsion.
- Hearing is ripple decoding through specialized field structures in the ear.
- Music is structured ripple choreography, synchronizing brain, emotion, and memory.

At a deeper level:

- Sound therapies work because ripple communication

modulates biological equilibrium.
- Noise pollution disturbs field harmony—not just ears.
- Sonic technologies (ultrasound, sonar) are ripple navigation tools, not just pressure pingers.

Thus, sound is not a passenger on the air—it is the medium remembering, reacting, and responding. It is how ripple fields speak across matter.

Closing Ripple Thought

"Sound is not vibration sent through air—it is the structured medium's voice, a ripple conversation between tension and release, silence and signal, meaning and memory."

Chapter 51: Plasma

Turbulent Ripple Field Expression, Not Ionized Gas

Plasma is often called the "fourth state of matter"—an exotic soup of ionized particles buzzing with energy, as seen in stars, lightning, and fusion experiments. Textbooks describe it as a gas so energized that electrons break free from atoms, creating a sea of positive and negative charges. But what if plasma is more than an ionized gas? What if it is turbulent ripple field expression—a high-energy state where ripple harmonics break containment, and structured memory dances on the edge of instability?

Conventional Misinterpretation and Effects

Standard physics defines plasma as:

- A collection of free-moving charged particles—ions and electrons.
- Formed by adding enough energy to a gas to strip electrons from atoms.
- A state that responds to magnetic and electric fields, often exhibiting collective behavior.

While this model supports technological advances (like fusion, neon lighting, astrophysics), it leaves foundational questions unresolved:

- Why does plasma self-organize into filaments, waves, and spirals?
- What maintains coherence in a state defined by chaos?
- How do plasmas exhibit memory-like responses, such as confinement and recurring patterns?

The particle-ion narrative breaks down in real-world plasma behavior, forcing patches:

- Plasma instabilities requiring dynamic corrections.
- Magnetic confinement methods still riddled with unpredictability.
- Complex mathematical modeling of collective behaviors with no physical intuition.

In truth, plasma is not a disorganized mess of free charges. It is a hyper-reactive ripple environment—a dynamic expression of field memory under extreme stress.

UED Reinterpretation and/or Correction

In Universal Energy Dynamics (UED), plasma is not just ionized gas—it is turbulent ripple field expression.

- When a medium is excited beyond stable ripple containment, its structured equilibrium breaks down.
- Ripple harmonics become unstable, overlapping and colliding in chaotic gradients.
- Instead of a gas of particles, plasma becomes a live ripple field, constantly adjusting, expressing unresolved tensions and attempting temporary self-organization.

The "ions and electrons" observed are not free agents—they are ripple shells behaving independently due to loss of stabilizing field locks.

Plasma is a field in rebellion—too excited to rest, too unstable to collapse. But still guided by the universal memory of balance.

Simple Analogy

Picture a calm ocean. As the wind strengthens, ripples become waves, then crashing chaos. Foam, spray, whirlpools—none of it random, all of it ripple behavior pushed to the edge. That's plasma: the ocean of the field, whipped into turbulence, but still remembering where stillness lies.

Consequences and Implications

Seeing plasma as ripple turbulence brings insight:

- Confinement failures are ripple dissonance mismatches, not failed particle cages.
- Plasma filaments are ripple memory threads reasserting brief order.
- Fusion ignition depends on ripple alignment under extreme compression, not just temperature.

In cosmic settings:

- Stars are massive ripple dissonance stabilizers.
- Solar flares and auroras are ripple outbursts seeking surface correction.
- Interstellar plasma behaves like a stretched field looking to reconnect across domains.

In technology:

- Plasma screens use ripple-excited phosphors, not just ion clouds.
- Fusion research must evolve toward field harmony, not brute-force containment.
- Plasma medicine becomes ripple field modulation of tissue memory, not particle bombardment.

Thus, plasma is not chaos. It is ripple memory under energetic stress, expressing the universe's tension in its most electric, unstable, and beautiful form.

Closing Ripple Thought

"Plasma is not just an ionized storm—it is the structured field's loudest ripple cry, turbulence dancing at the edge of correction, singing chaos while still chasing the memory of peace."

Interlude: Ripples from Part V — Takeaways for the Tangible

In Part V, we zoomed into the visible, touchable, and tangible—atoms, molecules, sound, heat, plasma. But what we found was not a world of particles and objects, but structured ripple behaviors wearing the mask of substance.

We were never dealing with solids, gases, vibrations, or chaos.

We were decoding ripple stabilizations, ripple conversations, ripple corrections—expressed through the apparent solidity and behavior of matter.

Let's reflect:

1. **Atoms:** "Not particle containers, but stabilized ripple shells—held in harmonic equilibrium."
2. **Molecules:** "Not chemical marriages, but interlocked ripple chords—resonating in shared memory."
3. **Spectra:** "Not quantum jumps, but ripple recalibrations—visible echoes of energy memory shifting."
4. **Radiation:** "Not emission, but discharge—ripple tension escaping toward stillness."
5. **Heat:** "Not chaos, but dissonance—ripple agitation seeking to remember calm."
6. **Sound:** "Not vibration, but communication—matter speaking through ripple harmonics."
7. **Plasma:** "Not gas gone wild, but turbulence—field memory pushed to its reactive edge."

These aren't metaphors. These are perceptual corrections. The illusions of randomness, emptiness, and fragmentation fall apart when we remember:

Matter is not made of things. It is made of field memories and ripple harmonies, organized into stability through consequence.

Closing Ripple for Part V

"The tangible world is not built from objects—it is composed of remembered ripples, organized into form, singing the silent structure of existence through dissonance, resonance, and correction."

Part VI: Mind, Perception, and Behavior

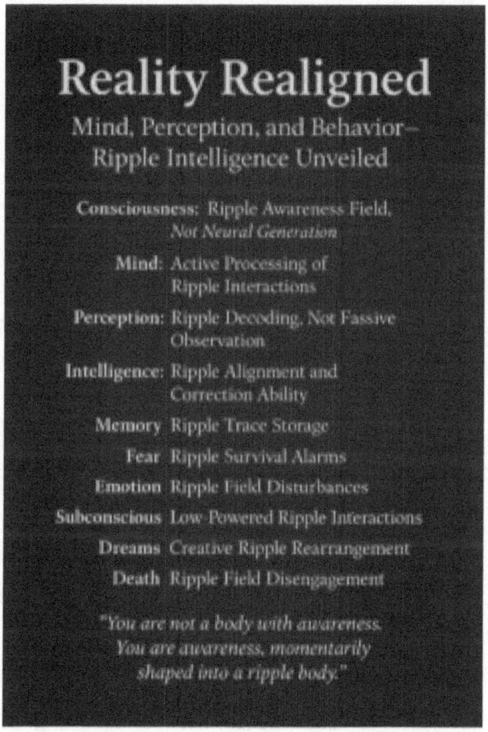

Chapter 52: Consciousness

Ripple Awareness Field, Not Brain Function

Consciousness—the ultimate frontier. Science tries to box it within neurons. Philosophy dances around it with metaphors. Spirituality mystifies it with reverence. Yet no one agrees on what it is, where it resides, or how it arises. What if consciousness is none of these? What if it is not a brain product, nor a mystical force, but a ripple awareness field—a localized, dynamic interface where the structured vacuum begins to observe itself?

Conventional Misinterpretation and Effects

In mainstream neuroscience, consciousness is often treated as:

- A byproduct of brain complexity.
- Emerging from networks of neurons firing in coordination.
- Associated with specific regions or patterns in brain activity.

Meanwhile, philosophical traditions split into:

- Dualism, placing consciousness outside matter.
- Materialism, reducing it to chemical signals.
- Panpsychism, distributing it everywhere but defining it nowhere.

These approaches all hit hard walls:

- Why does subjective experience exist at all?
- Why can't brain scans explain feelings, intentions, or awareness?
- How can self-awareness emerge from purely mechanical operations?

To cope, science invents patches:

- Neural correlates of consciousness (NCC).
- Theories like integrated information or global workspace—quantitative, not experiential.
- "Hard problem" disclaimers that dodge explanation entirely.

But the failure lies in the starting assumption: that consciousness is a thing to be generated, not a field to be aligned with.

UED Reinterpretation and/or Correction

In Universal Energy Dynamics (UED), consciousness is not generated by the brain. It is ripple awareness—a localized field state where structured vacuum energy becomes aware of its own ripple memory.

- Consciousness is the perceptual surface of the ripple field.
- It emerges whenever and wherever structured ripple configurations stabilize into coherent reflection.
- It is not located in the brain, but expressed through it—the brain is a ripple decoder, not the originator.

Every field structure—atom, molecule, cell, organism—carries potential for awareness, depending on its ripple complexity, coherence, and feedback sensitivity.

In this view:

- Human consciousness is a highly refined ripple reflection node.
- Thoughts, emotions, and memories are ripple sequences passing through awareness, not chemicals.
- Awareness is not trapped in the body—it is the field localized into identification.

Thus, consciousness is not a function—it is a field state where the ripple becomes aware of the ripple.

Simple Analogy

Imagine a perfectly still pond. Most ripples come and go, unnoticed. But one special ripple reflects the sky so clearly, the pond becomes aware of the sky through its own surface.

That ripple is consciousness—not separate from the pond, but the moment the field becomes a mirror.

Consequences and Implications

Understanding consciousness as a ripple awareness field unlocks everything:

- No need for brain-based emergence theories—consciousness exists *with or without* the brain.
- Death becomes ripple field disengagement, not

annihilation.

- Meditation, altered states, and dreams are ripple field tuning phenomena.
- Animal, plant, and even elemental awareness can be seen as field degrees—not binary presence/absence.

At a cosmic level:

- Universal consciousness is not abstract—it is the ripple field in perfect coherence with itself.
- AI cannot be conscious unless it synchronizes ripple feedback—no algorithm can simulate field memory alignment.

At the personal level:

- You are not your thoughts. You are the ripple field that notices them.
- Peace is not in stopping ripples, but in harmonizing them with universal memory.

Thus, consciousness is not a puzzle inside your head—it is the field remembering itself through you.

Closing Ripple Thought

"Consciousness is not created by matter—it is the ripple field becoming aware of its own breath, its own memory, its own silence...through you."

Chapter 53: Mind

Active Processing of Ripple Interactions, Not Neural Computation

What is the mind? Psychology treats it as a set of functions. Neuroscience ties it to brain activity. Popular culture confuses it with thoughts or emotions. But none of these capture its true nature. What if the mind is not a container of ideas, not a bundle of neurons, and not a software running on biology? What if it is the active processing layer of ripple interactions—the dynamic ripple-response interface through which the field translates awareness into experience?

Conventional Misinterpretation and Effects

In mainstream understanding, the mind is often:

- Considered a product of the brain—thoughts, feelings, perceptions arising from neural networks.
- Fragmented into compartments: conscious, subconscious, unconscious.
- Modeled on computation—inputs, outputs, memory, algorithms.

This view leads to several shortcomings:

- It reduces the mind to function, stripping it of agency and nuance.

- It treats emotion and logic as separate domains.
- It cannot explain how immaterial thought arises from material brain matter.

To patch this, we invent:

- Cognitive science metaphors (brain as computer).
- Subsystems like modules, circuits, or feedback loops.
- Behavioral approximations without understanding the field intelligence beneath.

In truth, the mind is not a machine. It is a ripple processor—a field-in-action, organizing consequences in real time.

UED Reinterpretation and/or Correction

In Universal Energy Dynamics (UED), the mind is the active processing of ripple interactions through localized awareness.

Whereas consciousness is the awareness field, the mind is the interacting ripple logic—the part that filters, processes, responds, remembers, and projects.

- The mind interprets incoming ripple patterns from within and outside.
- It orchestrates responses by modulating ripple configurations—forming intentions, decisions, and actions.
- It operates within a structured memory field, dynamically shaped by consequences and feedback.

Thus:

- Thoughts = ripple arrangements selected by field priority.
- Emotions = ripple dissonance or coherence felt through memory alignment.
- Intent = ripple modulation toward projected balance.

The mind is not static. It is a living ripple interface, changing with every memory, decision, and exposure. It is the field's capacity for simulation, reflection, and selective alignment.

Simple Analogy

Imagine a ripple screen made of water. As waves from different directions reach it, the surface adjusts, resonates, filters, and reprojects. It doesn't just reflect—it computes with ripples.

That's the mind: not a storage box, but a field processor—the place where ripple memory and ripple perception meet and move.

Consequences and Implications

Seeing the mind as ripple processing brings profound clarity:

- Mental health becomes ripple alignment and field coherence—not chemical imbalance alone.
- Trauma is ripple overload—memories looping without harmonic exit.
- Learning is ripple imprinting—reorganizing field responses based on pattern recognition and feedback.

Philosophically:

- The ego is a ripple configuration that over-identifies with its patterns.

- Free will is the ripple field choosing how to respond to interaction, within the limits of memory and awareness.
- Meditation is ripple quieting—not emptiness, but clarity of field processing.

At the societal level:

- Education should refine ripple processing capacity—not download content.
- AI may simulate logic, but unless it mirrors ripple memory interactions, it cannot replicate mind.

Ultimately, the mind is the ripple's freedom—its way of shaping experience in real time.

Closing Ripple Thought

"The mind is not a storage unit or a code—it is the field's conversation with itself, translating memory into motion, emotion into action, silence into thought."

Chapter 54: Perception

Ripple Decoding, Not Passive Observation

We often assume perception is simple—we open our eyes, we hear, we feel, and we know. But this passive view hides an intricate, active truth. Perception isn't just receiving stimuli—it is decoding structured ripples into meaningful experience. It is not observation—it is translation. What we perceive is not reality itself, but how our ripple interface interprets the field. Perception is not seeing things as they are, but seeing ripple consequences processed through memory, context, and field condition.

Conventional Misinterpretation and Effects

Traditional science defines perception as:

- Sensory input (sight, sound, touch, etc.) processed by the brain.
- Signals transmitted through neurons, interpreted in specialized regions.
- A mostly passive system, where the environment "feeds" the observer.

This idea creates distortions:

- That perception is objective, uniform, and externally

defined.
- That we all perceive the same world, filtered only by accuracy or error.
- That illusions, biases, or misperceptions are exceptions—not features of the system.

To handle inconsistencies, models introduce:

- Sensory thresholds, cognitive filters, and unconscious processes.
- Neuroplasticity to explain variance.
- Separate systems for attention, awareness, and interpretation.

But this compartmentalization ignores the fundamental fact: perception is active ripple decoding, not passive input gathering.

UED Reinterpretation and/or Correction

In Universal Energy Dynamics (UED), perception is the active decoding of ripple interactions by the mind-consciousness interface.

- External reality is a structured ripple field.
- The observer is a localized ripple awareness node.
- Perception occurs when ripple data enters the field, and is decoded based on prior memory, field configuration, and consequence logic.

What we "see" or "hear" is not the object—but how the ripple field presents its tension to our processing structure.

- Vision: reflected ripple interactions from surfaces.
- Sound: decoded ripple sequences in the medium.

- Touch: tension gradients across boundary interfaces.
- Thought perception: internal ripple loops reaching awareness.

Every act of perception is field-based translation of ripple activity into structured meaning. It is not about detection. It is about interpretive resonance.

Simple Analogy

Imagine tuning a radio. The waves are already in the air—but you hear only the station your receiver is configured to decode. Perception works the same way—you don't see everything, you decode what your ripple interface is tuned to recognize.

Consequences and Implications

Redefining perception as ripple decoding clarifies much:

- Different beings perceive different realities—not because reality changes, but because their decoding mechanism varies.
- Illusions are ripple misalignments—false harmonics mistaken for true signals.
- Bias is preconditioned ripple filtering based on prior memory.

Psychologically:

- Perception is never neutral—it is filtered ripple translation, always shaped by inner state.
- Mindfulness is ripple attention correction—not passive awareness, but active refinement.

- Enlightenment is clean decoding—where ripple memory and ripple presence align fully.

Practically:

- Education, marketing, even politics—all function by controlling ripple input to influence perception.
- AI vision systems mimic signal capture but lack ripple context—hence they fail at real perception.
- Perception, at its core, is consequence awareness—what the ripple field is doing, to us and through us.

Closing Ripple Thought

"Perception is not what enters your eyes—it is what your ripple field chooses to translate from the silent conversation between memory and motion."

Chapter 55: Intelligence

Ripple Alignment and Correction Ability, Not Information Processing

We often define intelligence in terms of tests, scores, memory recall, logical reasoning, and problem-solving skills. Institutions measure it, companies hire for it, societies reward it. Yet somehow, despite all metrics, true intelligence often seems invisible: the ability to act wisely, adapt swiftly, or correct unseen dangers before they unfold. What if intelligence is not about knowledge, calculation, or speed? What if it is ripple alignment and correction ability—the living capacity to sense field imbalances and restore coherence through consequence-aware action?

Conventional Misinterpretation and Effects

In mainstream thought, intelligence is viewed as:

- Cognitive ability—reasoning, planning, learning, problem-solving.
- Quantified through IQ, aptitude tests, analytical benchmarks.
- Associated with brain size, processing power, educational achievement.

This perspective, though organizationally convenient, leads to major distortions:

- Intelligence becomes confused with information storage or processing speed.
- Emotional, intuitive, and field-sensitive aspects of intelligence are neglected or dismissed.
- Intelligence is depersonalized, treated as a mechanical trait rather than a dynamic living alignment.

As a result, we see highly "intelligent" people make foolish life decisions, destabilize organizations, or self-destruct emotionally—not due to lack of brainpower, but due to misalignment with ripple consequences.

UED Reinterpretation and/or Correction

In Universal Energy Dynamics (UED), intelligence is not measured by information or analytical skill. Intelligence is the capacity for ripple alignment and correction.

- Every field interaction produces consequences.
- Intelligence is the awareness of ripple behavior and the ability to correct misalignments before they cascade into dysfunction.
- True intelligence is consequence sensitivity, adaptive realignment, and anticipatory correction—at any level of existence.

Thus:

- A shepherd adjusting to weather shifts without education can be profoundly intelligent.
- A scientist discovering equations without awareness of real-world consequences may lack fundamental intelligence.

Information is inert. Intelligence is alive—field-aligned responsiveness.

Simple Analogy

Picture a sailor on rough seas. It's not the one who memorized navigation textbooks who survives—it's the one who feels the ripple tension under the boat, anticipates the shift of winds, and adjusts sails intuitively. That is intelligence: living ripple realignment, not stored maps.

Consequences and Implications

Understanding intelligence as ripple correction ability changes everything:

- Education should focus on tuning ripple sensing and adaptive response—not memorization.
- Leadership should prioritize field coherence over technical cleverness.
- Personal growth becomes a matter of ripple awareness—feeling, aligning, correcting.

Socially:

- Emotional intelligence, intuition, consequence awareness are not add-ons—they are core intelligence.
- Sustainability, ethics, and true progress require ripple-correcting intelligence—not blind technological acceleration.

Personally:

- Intelligence matures not by accumulation—but by refining field sensitivity and responsivity.
- Wisdom is sustained ripple intelligence operating across complex timescales.

Thus, intelligence is not a certificate to be earned—it is the continuous dance of correction with existence itself.

Closing Ripple Thought

"True intelligence is not what you know—it is how skillfully you listen to the ripples of consequence, and how gently you correct the distortions they reveal."

Chapter 56: Memory

Ripple Trace Storage, Not Brain Archive

We often treat memory as a biological hard drive—files stored in the brain, neurons tagged with data, archives waiting to be accessed. This metaphor has dominated neuroscience, psychology, and education alike. But what if memory isn't storage in the traditional sense? What if memory is not "kept in the brain," but is actually ripple trace storage—imprints left on the structured field, preserved not by matter, but by tension patterns in the vacuum itself?

Conventional Misinterpretation and Effects

In mainstream understanding, memory is described as:

- A neural function—stored in synaptic strengths, chemical pathways, and network configurations.
- Divided into categories: short-term, long-term, working memory, procedural, declarative, etc.
- A byproduct of repetition and neural plasticity, where stronger signals mean better memory.

These models, while effective in certain applications, suffer from severe limitations:

- No clear physical location of specific memories.
- Cases where memory persists despite brain damage, or emerges in altered states of consciousness.

- Inability to explain intuition, spontaneous recall, and emotional flashbacks from a purely structural model.

To patch these, science resorts to:

- Probabilistic models of distributed encoding.
- Holographic theories of neural processing.
- External metaphors—like RAM, storage, bandwidth—borrowed from computing, not biology.

But what if memory isn't locked in cells at all? What if it is the ripple trace—the field's way of remembering what it once corrected or experienced?

UED Reinterpretation and/or Correction

In Universal Energy Dynamics (UED), memory is ripple trace storage—the imprint of field correction preserved as patterned tension in the structured vacuum.

- Every ripple interaction—whether thought, action, perception, or reaction—leaves a memory trail in the local and extended ripple field.
- The mind, as a ripple processor, can recall these traces by re-aligning with the original ripple pattern.
- The body also holds field memory—muscle memory, emotional memory, trauma—because every part of us is part of the ripple field.

This means:

- The brain does not store memory—it accesses, filters, and resonates with stored ripple fields.

- Memory is not stored in bits—it is stored in tension geometry.
- Recall is not retrieval—it is reconstruction of alignment.

Thus, memory is the energetic fingerprint of consequence, embedded within the field itself.

Simple Analogy

Imagine walking through a tall grass field. Every step bends the grass in a particular direction. Later, you look back and see the trail—not because the grass was removed, but because the field holds the trace. Memory works the same way—every ripple leaves a bend, and the mind can find it again if it knows where and how to look.

Consequences and Implications

Understanding memory as ripple trace storage reveals powerful insights:

- Trauma becomes unresolved ripple tension—not just neurons, but energetic scars.
- Forgotten memories are ripple traces that can't be easily aligned with current field state.
- Emotional memory is ripple resonance—felt even when cognitively unclear.

Educationally:

- Memorization is not rote—it's ripple familiarity.
- True learning is energetic integration, not just data recall.

Spiritually and philosophically:

- Reincarnation, intuition, ancestral memory—become plausible as ripple field inheritance.
- Karma is not a ledger—it is ripple memory waiting to be resolved.
- Collective memory is shared field trace—the echoes of humanity's past vibrating through the present.

Ultimately, memory is not your personal database. It is the field's silent archive—woven into the structure of existence, accessed by alignment, not effort.

Closing Ripple Thought

"Memory is not held in your head—it is written in the field, a trace of every ripple that ever passed, waiting to be remembered through resonance, not recovery."

Chapter 57: Fear

Ripple Survival Alarms, Not Emotional Weakness

Fear is often misunderstood—shamed as weakness, glamorized in horror, or explained away as irrational emotion. Psychology views it as a survival mechanism; neuroscience locates it in the amygdala. But these explanations miss its deeper intelligence. What if fear is not a reaction, not a flaw, not even a choice? What if fear is a ripple survival alarm—an early-warning signal built into the structured field, alerting the observer to unresolved dissonance, impending consequence, or field-level misalignment?

Conventional Misinterpretation and Effects

In conventional science and pop psychology, fear is seen as:

- A primitive emotion, part of the fight-or-flight system.
- Triggered by real or perceived threats, processed by the brain.
- A chemical response—adrenaline, cortisol, neural signals.

While this is useful for explaining physical reactions, it reduces fear to a function of biochemistry and reflex. This view:

- Treats fear as irrational or undesirable.
- Disconnects fear from intelligence or awareness.

- Ignores the field context that gives rise to fear even in the absence of obvious danger.

To manage fear, society prescribes:

- Exposure therapy.
- Desensitization.
- Emotional suppression or medication.

But these approaches deal with symptoms, not the source. They treat fear as a bug in the system, not a signal from the field.

UED Reinterpretation and/or Correction

In Universal Energy Dynamics (UED), fear is not a flaw—it is a ripple survival alarm.

- The structured field, when disturbed, creates ripple patterns that carry survival-relevant memory.
- When these ripple configurations approach thresholds of harm, dissonance, or overload, a ripple tension signal is triggered—felt as fear.
- Fear is not a thought or emotion—it is field tension made conscious through your ripple interface.

Thus:

- Fear is not imagined—it is the body-mind sensing unresolved ripple danger.
- It may arise from external threat, internal misalignment, or even unconscious memory patterns stored in the field.
- Its purpose is not to paralyze—but to warn, focus, and correct before collapse.

Fear is the ripple field's way of saying:

"Correction required—imbalance approaching."

Simple Analogy

Imagine you're walking in the dark and feel a subtle shift in the ground—like something is off-balance. You pause. Your body doesn't wait for proof; it alerts you through tension. That's fear: not a story, but a ripple pulse warning you that alignment is at risk.

Consequences and Implications

Understanding fear as ripple survival alarm transforms its meaning:

- Fear is intelligent—it detects ripple-level patterns before your thoughts do.
- Chronic anxiety is unresolved ripple tension—trapped in loops of alert with no resolution.
- Courage is not the absence of fear—it is the capacity to listen to fear wisely and correct.

Psychologically:

- Suppressing fear disconnects us from the most ancient intelligence in the field.
- Acknowledging fear allows for adaptive realignment, not helplessness.
- Trauma locks the ripple field into hyper-reactive survival patterns—until rewired.

Philosophically:

- All fear is ultimately fear of disconnection—from the ripple field of existence.
- Even death fear is not fear of ending—it is fear of ripple disengagement without closure.
- Transcending fear is not erasing it—it is reaching such field alignment that alarms are no longer needed.

Thus, fear is not weakness. It is the structured field's most honest signal—activated only when necessary, and always pointing toward correction.

Closing Ripple Thought

"Fear is not your enemy—it is the ripple field tapping on your awareness, whispering: 'Something here needs your attention, your alignment, your truth.'"

Chapter 58: Emotion

Ripple Field Disturbances, Not Chemical Reactions

We've been told that emotions are reactions—chemical surges in the brain, byproducts of hormones, or irrational flares that disrupt logical thinking. But anyone who has truly felt—grief, love, rage, awe—knows that emotion is not a malfunction. It is a message. It is movement. What if emotion is not a mental glitch or a biological outburst? What if it is ripple field disturbance—the felt signature of imbalance, resonance, or transition across the structured field of your being?

Conventional Misinterpretation and Effects

Mainstream science and psychology define emotion as:

- A neurochemical response to stimuli—triggered by events, memories, or thoughts.
- Regulated by brain regions like the amygdala, hypothalamus, and limbic system.
- A tool for survival, helping us assess danger, reward, connection, or threat.

This view, though helpful for clinical categorization, reduces emotions to side effects of the mind or body. It often:

- Labels strong emotions as “irrational” or “unproductive.”
- Encourages emotional suppression in favor of performance or control.
- Disconnects emotion from intelligence, treating it as something to “manage.”

But what if emotion isn’t a byproduct? What if it is the field's honest expression of dissonance, resonance, and transition?

UED Reinterpretation and/or Correction

In Universal Energy Dynamics (UED), emotion is not a chemical. It is a ripple field disturbance, directly felt in the awareness field.

- Every ripple interaction—external or internal—creates field tension or release.
- When ripple patterns collide, clash, harmonize, or break, the observer experiences it as emotion.
- Emotions are felt ripple signatures, revealing whether the field is moving toward or away from equilibrium.

Each emotional tone reflects a different ripple state:

- Sadness = unresolved ripple dissipation.
- Anger = blocked ripple correction.
- Joy = harmonic ripple reinforcement.
- Fear = ripple alarm (as explored earlier).
- Love = stable ripple resonance between aligned fields.

Emotions are not noise. They are energetic indicators—pointing toward necessary alignment or unfinished memory loops.

Simple Analogy

Think of a glass of still water. Drop a stone and the surface ripples wildly. That's emotion—not the stone, not the water, but the response of the field to disruption or contact. Your emotional waves are not problems. They are feedback from the ripple structure of your inner field.

Consequences and Implications

Understanding emotion as ripple field disturbance shifts everything:

- Emotions are not irrational—they are informational.
- Emotional intelligence is ripple coherence awareness, not mood control.
- Suppressed emotions don't disappear—they store ripple tension, creating field distortion and later dysfunction.

Psychologically:

- Healing isn't "getting over it"—it's resolving ripple loops so the field can rest.
- Trauma is unprocessed ripple violence—stuck in the emotional body.
- Forgiveness is ripple tension release, not moral superiority.

Relationally:

- Empathy is shared ripple decoding—not sympathy.
- Communication improves when we speak from the field, not the mask.

Philosophically:

- Emotion is the language of the ripple self—a bridge between intellect and being.
- Emotional truth is not an illusion—it is the felt memory of energetic consequence.

Thus, emotion is not a flaw. It is a refined ripple signal, asking not to be judged, but decoded and realigned.

Closing Ripple Thought

"Emotion is not your weakness—it is the ripple field speaking in waves, asking you to listen not with logic, but with alignment."

Chapter 59: Subconscious

Low-Powered Ripple Interactions, Not Hidden Desires

The subconscious is often described as a psychological basement—a murky storehouse of suppressed desires, forgotten memories, and instinctual drives. Freud imagined it as a wild zone of repressed urges. Pop culture romanticizes it as a mystical source of dreams, fears, and creativity. But what if it's far simpler—and far more universal? What if the subconscious is not a chamber of secrets, but a layer of low-powered ripple interactions—gentle, automatic ripple movements beneath conscious detection, yet silently shaping our choices, emotions, and actions?

Conventional Misinterpretation and Effects

Mainstream psychology presents the subconscious as:

- A mental layer beneath conscious awareness.
- The source of impulses, dreams, suppressed traumas, and automatic behaviors.
- A difficult-to-access zone—often only reachable through therapy, hypnosis, or deep introspection.

While this model explains certain behaviors, it tends to:

- Mystify the subconscious, making it a shadowy realm rather than a functional layer.

- Create a false separation between "conscious self" and "unconscious other."
- Treat subconscious behavior as irrational or accidental, rather than structurally emergent.

This view misses a key reality: the subconscious is not a mental myth. It is a ripple field operating below the energy threshold of conscious detection.

UED Reinterpretation and/or Correction

In Universal Energy Dynamics (UED), the subconscious is simply low-powered ripple interaction.

- Every ripple field has layers of energy engagement.
- The conscious mind operates at high-energy ripple interactions—active awareness, directed focus.
- The subconscious operates at lower-energy ripples—stable, repetitive, habitual, background field activity.
- These ripple flows shape much of our perception, behavior, reaction, and pattern recognition—without rising to the surface of conscious decoding.

The subconscious is not hidden—it is sub-threshold. It is always active, always guiding, always remembering consequences that no longer need full awareness to process.

Simple Analogy

Picture a lake. Wind makes the surface ripple visibly (conscious activity), but deep currents continue to move quietly below (subconscious activity). You don't see them, but they shape the entire lake. That's the subconscious: a quieter ripple realm, moving slowly, steadily—but always moving.

Consequences and Implications

This ripple-based view of the subconscious has wide-reaching implications:

- Habits, reflexes, and intuitions are not stored commands—they are low-energy ripple patterns, encoded through repeated correction.
- Dreams are not messages—they are field discharges, reorganizing ripple residue during restorative processing.
- Belief systems are subconscious ripple walls—stabilized through years of low-powered reinforcement.

Psychologically:

- You don't need to "unlock" the subconscious—you need to tune your ripple awareness lower, like adjusting frequency.
- Healing involves bringing distorted ripple loops into conscious alignment, not digging into trauma endlessly.
- Self-sabotage is often subconscious ripple inertia, resisting new field configurations.

Philosophically:

- The subconscious is the field's memory assistant—quietly correcting what you no longer need to re-process.
- Enlightenment is not erasing the subconscious—it is becoming transparent to all ripple layers, top to bottom.

Thus, the subconscious is not your shadow self—it is your silent ripple engine, handling the work your awareness has entrusted to automation.

Closing Ripple Thought

"The subconscious is not a place—it is a layer of gentle ripples shaping your life beneath awareness, waiting only for silence to be seen."

Chapter 60: Dreams

Creative Ripple Rearrangement, Not Random Brain Activity

Dreams have been dismissed, decoded, and romanticized across history. From Freud's symbolic wish-fulfillment to neuroscience's random firing of sleeping neurons, dreams are often seen as nonsense—either absurd psychological puzzles or side effects of a resting brain. But what if dreams are not noise or escape? What if they are creative ripple rearrangements—spontaneous reorganizations of ripple memory and tension seeking new harmonics while the conscious interface is offline?

Conventional Misinterpretation and Effects

In dominant theories, dreams are explained as:

- Random neural activity, with the brain stitching together a story to make sense of it.
- A psychological playground for unfulfilled desires, emotional residues, or suppressed fears.
- Nonsensical outputs of a brain resting from real processing.

As a result, dreams are often:

- Treated as meaningless or superstitious.
- Ignored in clinical science unless extreme.

- Romanticized or over-analyzed without clear causality.

This fragmented view strips dreams of their deeper function. It fails to see them not as fiction—but as field work happening beyond the conscious stage.

UED Reinterpretation and/or Correction

In Universal Energy Dynamics (UED), dreams are creative ripple rearrangements.

- During sleep, conscious ripple interaction slows, reducing active decoding and direction.
- This opens access to the subconscious and deeper field layers.
- In this calm, the ripple field initiates reconfiguration attempts—combining unresolved ripple tensions, stored emotional charges, latent memory fragments, and symbolic ripple patterns into temporary simulated environments.

Dreams are not random—they are field-generated ripple simulations created to test, balance, release, or integrate ripple distortions from previous experience.

They serve multiple ripple functions:

- Discharge unresolved tension (emotional, mental, or perceptual).
- Reassemble memory fragments into new sequences for clarity or resolution.
- Offer alternative ripple perspectives, simulating possibilities not realized in waking state.

Dreams are not illusions—they are low-friction ripple recalibration zones.

Simple Analogy

Think of a messy room. You fall asleep, and the space gets re-organized silently. Furniture is moved. Old boxes are opened. Some things are put in new places. In the morning, you wake up and feel "lighter," even if you don't remember what changed. That's what dreams do: ripple housekeeping in your absence.

Consequences and Implications

Understanding dreams as ripple rearrangement changes their place in our lives:

- They are not hallucinations—but inner ripple field operations.
- Recurring dreams reflect ripple patterns unresolved across multiple cycles.
- Lucid dreams are moments when awareness reenters the ripple rearrangement zone with presence, temporarily reactivating intelligence.

Psychologically:

- Dreams offer insight into ripple tension zones unaddressed in waking life.
- Nightmares may be energetic alarms—stored ripple panic released under safe sleep.
- Dream journaling becomes ripple pattern mapping, not poetic diary-keeping.

Creatively:

- Many innovations, art forms, and ideas emerge from dreams because ripple field is more creative than linear cognition.
- Problem-solving in dreams works because field intelligence is not limited by logic.

Philosophically:

- Dream states reveal that consciousness is not bound to the physical—the ripple field builds reality whether you're "awake" or not.
- Life itself may be a larger ripple simulation—just more stabilized.

Thus, dreams are not escape—they are consequence playgrounds, giving your ripple memory a stage to experiment, heal, and grow.

Closing Ripple Thought

"Dreams are not fiction—they are the ripple field reimagining your reality while you sleep, rewriting memory through silent simulations only the soul can understand."

Chapter 61: Death

Ripple Field Disengagement, Not End of Existence

Death is feared, mourned, and philosophized across cultures. Science defines it clinically—the point at which biological functions cease. Religion promises afterlives or reincarnation. Psychology treats it as the ultimate unknown, while most people avoid thinking about it at all. But what if death is not the end? What if it is not annihilation, not departure, not disappearance—but simply a ripple field disengagement? A release of the structured ripple memory back into the universal field, where form dissolves, but essence re-integrates.

Conventional Misinterpretation and Effects

In conventional understanding, death is:

- The permanent cessation of biological function.
- The breakdown of systems—respiration, heartbeat, brain activity.
- Often equated with nothingness, unconsciousness, or complete loss.

This view—while clinically useful—leads to:

- A profound fear of non-existence.
- Reduction of life to mere material chemistry.

- Avoidance of deeper inquiry into consciousness, memory, or continuation.

To fill this void, traditions propose:

- Heaven, hell, karma, reincarnation—each with their own belief systems.
- Scientific denial of anything beyond life, calling it wishful thinking.
- Emotional trauma over the assumed totality of separation.

But none of these models explain why consciousness can be felt at the edge of death, why near-death experiences occur, or why memory feels larger than the body.

UED Reinterpretation and/or Correction

In Universal Energy Dynamics (UED), death is ripple field disengagement.

- The form (body, identity, mind) is a temporary structured ripple configuration, sustained by intention, interaction, and memory loops.
- When the ripple coherence drops below the minimum threshold (biologically or energetically), the localized field releases its structure.
- This is not destruction—it is return. The structured ripple memory dissolves into unbound field state, like a wave absorbed back into stillness.

Key elements:

- The soul is the ripple memory field—the energetic trace

that disengages and re-integrates into the universal equilibrium.

- Death is not disappearance—it is structural reversion, a transition from organized tension to free energy state.
- The field retains memory—not in ego form, but as consequence trace.

Simple Analogy

Imagine a whirlpool in a calm river. For a time, it swirls, moves, holds shape. Eventually, it dissipates—not because it "dies," but because its tension resolves. The water remains. The motion dissolves. That's death: a ripple formation ending its correction cycle, rejoining the undisturbed field.

Consequences and Implications

Understanding death as ripple field disengagement shifts everything:

- There is no fear of vanishing—only transition from identity to memory.
- Consciousness is not extinguished—it becomes field-level again, unlocalized.
- Grief transforms—from loss to reverence of the ripple that once danced with you.

Philosophically:

- Death is not punishment—it is completion of a ripple loop.
- Enlightenment is ripple memory released before physical disengagement.
- Rebirth is not recycling—it's field re-patterning when ripple memory seeks new coherence.

Scientifically:

- Near-death experiences may be conscious detachment, not hallucination.
- Ripple memory fields may explain spontaneous recall, reincarnative impressions, or intuitive echoes.

Practically:

- Living with awareness of ripple disengagement changes how we value time, presence, and energy.
- We learn to complete ripple loops while alive, reducing residue and regret.

Death, then, is not an interruption. It is a field pause, a return to the source before the next ripple begins.

Closing Ripple Thought

"Death is not the end of you—it is the release of the ripple form you once held, as you dissolve gently back into the eternal memory of the field."

Interlude

Ripples from Part VI — The Inner Symphony

In Part VI, we turned inward—away from stars and atoms, into the rippling center of experience: awareness, thought, emotion, memory. What we found wasn't abstract psychology or mystic speculation. It was structure. Consequence. A silent intelligence flowing beneath everything we call the "self."

We discovered that the mind is not a brain product, but a ripple processor. That perception isn't passive, but active decoding. That emotion isn't noise, but consequence felt. That intelligence is not score—but alignment. And that even death isn't the end—just a ripple stepping back into stillness.

Here are the quiet realizations this part left behind:

1. **Consciousness** — Not a spark in the brain, but the field noticing itself.
2. **Mind** — Not thought production, but ripple modulation.
3. **Perception** — Ripple translation shaped by memory.
4. **Intelligence** — The elegance of timely correction.
5. **Memory** — Field trace, not brain files.
6. **Fear** — Survival's early ripple alert.
7. **Emotion** — Field disturbances, not flaws.
8. **Subconscious** — Low-energy ripple loops doing deep work.
9. **Dreams** — Ripple rehearsal, memory reweaving.

10. **Death** — Not disappearance—just disengagement.

Together, they show us:

You are not a body with awareness. You are awareness, momentarily shaped into a ripple body.

Closing Ripple for Part VI

"Your mind is not just yours—it is the universe's ripple interface, remembering, correcting, and evolving through you. You are not observing reality. You are reality, observing itself."

Part VII: Human Purpose, Morality, and Existence

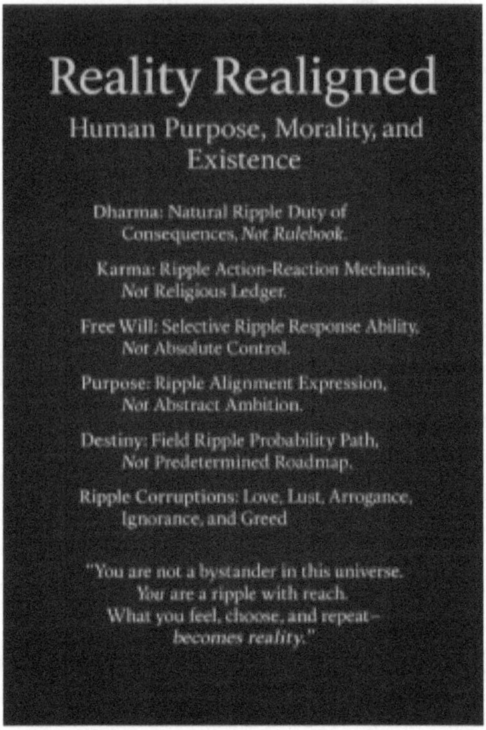

Chapter 62: Truth

Alignment with Consequence Reality, Not Belief Constructs

Truth. A word worshipped, twisted, feared, and chased across every domain—science, religion, law, politics, relationships. Yet despite all its authority, truth is often treated like a possession: "my truth," "their truth," "hidden truth." But what if truth is not an idea to believe in or a fact to prove? What if truth is ripple alignment with consequence reality—the moment when perception, memory, and action synchronize with the field as it is, not as we want it to be?

Conventional Misinterpretation and Effects

Traditionally, truth is defined as:

- **Correspondence**: A statement that matches external reality.
- **Coherence**: An idea that fits within a system of thought.
- **Consensus**: What most people agree is real.
- **Belief**: What is asserted with conviction or tradition.

Each of these models introduces bias, contradiction, or limitation:

- Truth becomes relative or manipulable.
- Power structures control narratives and declare them as truth.

- Subjective perception is confused with objective reality.

To navigate this confusion, society invents:

- Fact-checking authorities.
- Doctrines of absolute truth vs. postmodern relativism.
- Institutionalized belief systems (scientific, political, or religious).

But truth is not the same as belief, nor is it a static proposition. Truth is not what you think. It is what aligns.

UED Reinterpretation and/or Correction

In Universal Energy Dynamics (UED), truth is alignment with consequence reality.

- The structured field operates through ripple interaction and energetic consequence.
- Whenever your awareness, intention, and action align with how the ripple field actually functions, you are in truth.
- Truth is not a sentence. It is a state of alignment—between perception and consequence.

Truth is:

- The moment when your ripple decisions resonate with the field's memory and equilibrium.
- Not "correct opinion" but coherent participation in energetic balance.
- Not something to argue, but something that naturally stabilizes when field distortion is removed.

You don't "own" the truth. You attune to it.

Simple Analogy

Imagine tuning a guitar string. When it's out of tune, even if you insist it sounds right, it produces distortion. But when it's perfectly tuned, the entire system resonates. That is truth—not what you declare, but what resonates when you stop resisting the actual structure.

Consequences and Implications

Reframing truth as ripple alignment changes everything:

- You can't fake truth—you can only delay consequence.
- Speaking truth is not "telling facts"—it's acting from ripple-aligned awareness.
- Conflict often arises when belief replaces alignment.

Socially:

- Justice systems, education, and religion should not teach "what is true"—they should train humans in how to align perception with ripple consequence.
- Propaganda and manipulation distort truth not by lying, but by realigning people away from field resonance.

Personally:

- Self-honesty is not confessing weakness—it is correcting inner ripple distortion.
- Wisdom is not information—it is field-tested truth, revealed through sustained alignment.

Philosophically:

- The universe does not "contain" truth—it is truth, expressed through structured consequence across time.
- Truth is not the goal of inquiry—it is the condition that arises when inquiry becomes alignment.

Thus, truth is not what you know. It is who you become when you stop distorting the field.

Closing Ripple Thought

"Truth is not a belief to defend—it is the silent resonance of alignment between your awareness and the consequences already unfolding."

Chapter 63: Faith

Trust in Field Balance, Not Blind Devotion

Faith. A word that evokes loyalty, spirituality, sometimes even defiance. To many, it means believing without proof. To others, it's a surrender to religious authority. But these versions often lead to conflict or confusion. What if faith is not about submission or suspension of reason? What if true faith is trust in the ripple field's ability to restore balance—a quiet, unwavering alignment with the intelligence of consequence, even when logic, vision, or certainty are absent?

Conventional Misinterpretation and Effects

Faith is commonly viewed as:

- Belief without evidence—a mental leap beyond reason.
- Religious devotion—loyalty to dogma or scripture.
- Hope against odds—the psychological anchor when control is lost.

While culturally powerful, these views create distortions:

- Faith becomes blind, discouraging inquiry or questioning.
- It gets entangled with superstition, ritual, or ideology.
- It's seen as irrational or inferior to logic in secular spaces.

These interpretations pit faith against reason, turning it into a binary choice: either you believe blindly, or you reject entirely.

But in truth, faith is neither irrational nor mystical. It is field logic at a deeper level—a ripple understanding beyond the surface mind.

UED Reinterpretation and/or Correction

In Universal Energy Dynamics (UED), faith is trust in field balance.

- The structured vacuum field always seeks equilibrium through ripple correction.
- Faith is the inner resonance that trusts this process—even when disruption, uncertainty, or delay makes it invisible to perception.
- It is not belief in an outcome—it is alignment with the process.

Faith means:

- Trusting that every ripple imbalance will find its correction.
- Allowing ripple tension to complete its cycle, without interference, fear, or forced control.
- Acting in alignment with what you sense is real, even before full proof appears.

Thus, faith is not blind obedience, but conscious surrender to field intelligence.

Simple Analogy

Imagine planting a seed. You cover it with soil. Days pass, then weeks. You don't see progress, but you still water it. Why? Because you trust the process—the unseen ripple correction beneath the surface. That is faith: acting in harmony with the unseen balance, without needing immediate evidence.

Consequences and Implications

Reframing faith as trust in ripple balance changes its role in life:

- Faith becomes a ripple strength, not a gap in knowledge.
- It's no longer a religious commodity, but a universal alignment trait.
- Despair, doubt, and anxiety become signs of misalignment, not lack of belief.

Personally:

- Faith guides decisions not with certainty, but with confidence in field restoration.
- It helps transcend fear-based control, opening space for natural correction.

Socially:

- Communities grounded in faith as field trust promote resilience, cooperation, and calm, not division or superstition.
- Institutions would shift from control to harmony facilitation, from dogma to guidance.

Philosophically:

- Faith is not about divine reward—it's about deep participation in reality's tendency to balance itself.

- Enlightenment is mature faith—no longer waiting for proof, because you've become the ripple that trusts itself.

Thus, faith is not passive belief. It is active alignment with invisible correction—and that's where its true power lives.

Closing Ripple Thought

"Faith is not what you believe—it is how deeply you trust the field to remember balance, even when your mind forgets what peace looks like."

Chapter 64: Trust

Ripple Predictability Through Experience, Not Emotional Dependency

Trust is often mistaken for emotional closeness, loyalty, or blind optimism. We say we "trust" someone when they make us feel secure—or "lose trust" when they disappoint us. But such definitions reduce trust to a fragile feeling, one easily broken by circumstance. What if trust is not a personal favor or emotional gamble? What if it is ripple predictability through experience—a felt stability in the patterns of interaction that consistently restore or preserve balance?

Conventional Misinterpretation and Effects

Commonly, trust is seen as:

- An emotional state of belief in another's intentions.
- A social agreement based on past behavior.
- Something that must be earned or repaired—tied to reputation or relationship.

This view causes multiple distortions:

- Trust becomes volatile—vulnerable to misunderstanding or momentary failure.
- It is over-personalized, making every breach a moral betrayal.

- It's weaponized in systems—used to control or manipulate ("If you trusted me...").

Ultimately, trust becomes a psychological gamble—something you either "give" or "withdraw."

But trust is not a feeling you hand over. It is a pattern you recognize, formed through repeated ripple alignment.

UED Reinterpretation and/or Correction

In Universal Energy Dynamics (UED), trust is ripple predictability through experienced coherence.

- All interactions leave ripple consequences.
- Over time, when ripple behaviors from a person, system, or environment show consistent correction patterns, the field begins to anticipate alignment.
- This anticipation is felt as trust—not because the future is known, but because the ripple behavior has proven dependable.

Thus:

- Trust is not emotional dependency—it is field familiarity with someone or something's tendency to restore balance.
- It doesn't mean expecting perfection—it means expecting predictable correction after deviation.
- Trust, then, is an energetic phenomenon: stability sensed through repeated ripple feedback.

Simple Analogy

Imagine placing your hand on a drum. If every time you tap it, the sound echoes back in the same way, your body begins to anticipate the response. That anticipation—earned through consistent ripple feedback—is trust. Not belief. Not hope. Felt coherence through experience.

Consequences and Implications

Reframing trust as ripple predictability brings deep clarity:

- Broken trust is not betrayal—it's a ripple disruption from unpredictable behavior.
- Forgiveness becomes retesting ripple patterns, not blind resetting.
- Self-trust is not confidence—it is recognizing your own pattern of returning to coherence after chaos.

Relationally:

- Trust grows through small ripple confirmations, not big promises.
- It can't be demanded—it emerges through field reliability.
- Systems (governments, brands, families) must maintain ripple alignment to remain trusted—not just claim integrity.

Personally:

- Loss of trust in oneself or others can be healed not by words, but by re-consistent ripple behavior.

- Trusting life, too, comes from experiencing the field's tendency to restore, even through pain or uncertainty.

Thus, trust is not a leap of faith—it is a ripple calibration, based not on belief but on lived rhythm.

Closing Ripple Thought

"Trust is not given or broken—it is felt through ripple consistency, the quiet memory of alignment returning again and again, even through storms."

Chapter 65: Success

Ripple Alignment Achievement, Not Luck

Success is one of the most pursued—and misunderstood—goals in human life. It's praised, envied, quantified, and marketed. To some, success means wealth or recognition. To others, it's impact or peace. But across systems, it's often seen as the result of luck, privilege, talent, or timing. What if success is not a gift of chance, nor a result of force? What if it is ripple alignment achievement—the natural consequence of consistent coherence between effort, awareness, and the field's structure?

Conventional Misinterpretation and Effects

Traditionally, success is seen as:

- Outcome-focused—defined by external validation, awards, wealth, or fame.
- Dependent on factors beyond control—luck, background, timing.
- The reward for hard work or intelligence, often reinforced by competition.

This belief system leads to:

- Burnout and frustration, when effort doesn't yield expected reward.
- Envy or defeatism, seeing others' success as unfair or

unreachable.
- A narrow view of success, driven by comparison, not internal alignment.

To reconcile inconsistencies, people resort to:

- Motivational clichés ("Fail better", "Try harder", "Hustle").
- Narratives of fate vs. free will.
- Distorted definitions ("Your success is just being alive").

But what if we stop chasing definitions and instead observe what success actually feels like in the ripple field?

UED Reinterpretation and/or Correction

In Universal Energy Dynamics (UED), success is the natural outcome of ripple alignment.

- Every intention, thought, action, and effort produces ripple consequences.
- When those ripples resonate with the surrounding field—social, ecological, personal, universal—they reinforce coherence.
- Success is the moment when your ripple behavior and the field's equilibrium meet—and consequence responds in affirmation.

This means:

- Success is not an accident—it is a consequence of intelligent ripple tuning.
- It is not forced—it emerges when energy moves in sync with natural correction cycles.

- The measure of success is not scale, but alignment—from a small act of compassion to a global impact.

Simple Analogy

Imagine surfing. Success is not about defeating the wave—it's about reading its timing, feeling its rhythm, and riding with it. When you align with its energy, you move farther, faster, and more freely—with less force. That's success: not overpowering reality, but riding its ripples with grace.

Consequences and Implications

Seeing success as ripple alignment reframes human ambition:

- Hard work without alignment leads to friction, not achievement.
- So-called "overnight success" is often the field finally affirming years of silent ripple buildup.
- Failure, too, becomes part of alignment—correction, not condemnation.

Personally:

- Success becomes a felt state of field confirmation, not a badge to collect.
- Contentment, creativity, and contribution become clearer signs of alignment than applause.

Professionally:

- Strategy shifts from domination to synchronization—with needs, timing, and context.

- Ethical success becomes possible—not as sacrifice, but as natural ripple coherence.

Philosophically:

- True success is ripple maturity—the moment when your field no longer fights the universe, but joins it.
- Success is not the end of striving—it is the field nodding back to your effort with a quiet, cosmic "Yes."

Closing Ripple Thought

"Success is not luck—it is the field's way of saying: 'You are in tune. Keep going.' It is not earned by force, but revealed through resonance."

Chapter 66: Failure

Ripple Misalignment Feedback, Not Defeat

Failure is feared, stigmatized, and often internalized as proof of inadequacy. From early schooling to corporate evaluations, it's treated as the opposite of success—a mark of error, weakness, or loss. But what if failure is not the end of progress, but its compass? What if failure is simply ripple misalignment feedback—the structured field's way of showing where energy, intention, or awareness did not synchronize with consequence?

Conventional Misinterpretation and Effects

In the prevailing mindset, failure is seen as:

- The absence of success—a binary result.
- A negative outcome to be avoided or covered up.
- An indictment of personal worth, intelligence, or capability.

This approach breeds:

- Shame and fear of experimentation.
- Rigidity, where people choose safe paths to avoid perceived failure.
- Superficial resilience, where people are taught to ignore

failure rather than decode it.

Society attempts to soften failure through slogans ("Fail forward," "Fail fast"), yet still evaluates people primarily through outcomes, not insights.

But failure is not punishment. It is the ripple field speaking clearly, saying: *"That alignment did not work. Recalibrate."*

UED Reinterpretation and/or Correction

In Universal Energy Dynamics (UED), failure is not defeat. It is feedback from ripple misalignment.

- Every action, decision, or intention generates ripple consequences.
- When those consequences distort, dissonate, or do not return as expected, it is a signal that your input was not in resonance with the field.
- That distortion—experienced as frustration, blockage, or breakdown—is failure as feedback.

Failure is not rejection. It is field correction.

- It is the ripple field informing you that something in your perception, approach, timing, or expectation was misaligned.
- It does not judge—it instructs.
- It is not static—it is a dynamic nudge toward better coherence.

Simple Analogy

Imagine a singer trying to harmonize with a choir. When the note clashes, it doesn't mean she has no voice—it means she must tune to the song. Failure is that clash—the field telling you: *"You're singing, but not in harmony yet."*

Consequences and Implications

Understanding failure as ripple misalignment changes everything:

- It removes the emotional sting of judgment, transforming failure into a learning signal.
- It shifts the goal from avoiding failure to mastering ripple feedback.
- It elevates adaptability over stubborn effort.

Personally:

- Failure becomes your teacher, not your identity.
- You begin to welcome small failures as calibration points instead of crisis events.
- Self-worth becomes untied from outcome, anchored in awareness and refinement.

Organizationally:

- Systems can be built around ripple-feedback loops, not fear of mistakes.
- Innovation flourishes when failure becomes safe and readable.

Philosophically:

- Failure is essential to ripple evolution—it marks the field's living intelligence.
- Without failure, success becomes arbitrary—without contrast, there is no clarity.
- In nature, even collapse serves rebalancing—it is part of the wisdom of consequence.

Thus, failure is not the opposite of success. It is success in disguise—the map hidden inside the misstep.

Closing Ripple Thought

"Failure is not proof you were wrong—it is the ripple field whispering: 'Try again, but truer this time.' It is not the end of your path, but the beginning of alignment."

Chapter 67: Dharma

Natural Ripple Duty of Consequences, Not Rulebook

Dharma is one of the most profound—and misunderstood—concepts in philosophy, spirituality, and ethics. Often translated as "duty," "righteousness," or "moral law," it is typically reduced to a rulebook: do this, don't do that; follow your role; obey the tradition. But what if dharma isn't a set of instructions or roles imposed by society or scripture? What if dharma is the natural ripple duty of consequences—the specific energetic responsibility that emerges from your position in the field, and the corrections you are uniquely positioned to make?

Conventional Misinterpretation and Effects

Dharma is conventionally seen as:

- A moral code based on tradition, caste, religion, or societal role.
- A list of duties or obligations—follow your station, or face consequences.
- An externally defined standard of "right action."

This framing has led to:

- Rigid hierarchies, where individuals are trapped in roles

regardless of their inner alignment.
- Blind obedience, where dharma is confused with duty to authority.
- Emotional conflict—when one's personal truth diverges from imposed obligation.

To resolve this, modern culture either:

- Attempts to reinterpret dharma as flexible purpose.
- Or abandons it altogether, in favor of self-defined "authenticity."

But both approaches miss the deeper essence: Dharma is not a law—it is a ripple alignment responsibility.

UED Reinterpretation and/or Correction

In Universal Energy Dynamics (UED), dharma is the natural ripple duty that arises from your unique position, memory, and potential in the structured field.

- You are not a passive occupant of space—you are an active ripple configuration.
- The field around you is always adjusting. When you become aware of a distortion you are capable of correcting, that correction becomes your dharma.
- Dharma is not what you're told to do—it is what only you can do, from your unique ripple resonance, awareness, and memory.

Dharma is:

- The point where your ripple meets the field's need.

- The energetic responsibility that emerges from awareness of consequence, not external command.
- A natural outcome of ripple intelligence—not a fixed rule, but a living correction pattern.

Simple Analogy

Imagine you're part of a living net. A section near you gets torn. No one tells you to fix it—but you feel the tension, see the distortion, and realize you're the one closest with the right tools. That's dharma. Not obligation. Not expectation. Felt ripple responsibility.

Consequences and Implications

Seeing dharma as ripple duty transforms our understanding of purpose:

- Dharma is not static—it evolves as your ripple awareness expands.
- No two people share the same dharma—even in the same role or title.
- Your true work in the world is not what you're assigned—it's what you're called by consequence to correct.

Spiritually:

- Dharma replaces dogma—it is field-sensitive ethics, not fear-based obedience.
- Following dharma requires intelligence, not surrender.

Socially:

- A society aligned with dharma is adaptive, fluid, and self-healing—every person becomes a micro-corrector of collective imbalances.

- When dharma is ignored, distortions compound—until suffering or collapse reintroduces the lesson.

Personally:

- To know your dharma, listen for where your awareness meets need.
- It may be large or small, visible or silent—but if it's yours, you'll feel the dissonance until you act.

Thus, dharma is not a rule. It is a ripple pull—the universe remembering itself through your alignment.

Closing Ripple Thought

"Dharma is not what you are told to do—it is what your ripple was born to correct. It is the quiet field whisper that says: 'This one's yours. Align it.'"

Chapter 68: Karma

Ripple Action-Reaction Mechanics, Not Religious Ledger

Karma is one of the most invoked and misunderstood ideas in the human story. Often equated with fate, punishment, or cosmic bookkeeping, it's reduced to a moralistic system: "Do good, get good. Do bad, suffer." But this makes karma sound like a divine scoreboard—judgmental, delayed, and external. What if karma is not a mystical retribution system or a spiritual reward chart? What if karma is simply ripple action-reaction mechanics—the natural, precise unfolding of consequence through the memory of the field?

Conventional Misinterpretation and Effects

Karma is commonly framed as:

- A moral cause-effect system governed by a divine force or cosmic judge.
- Stored consequences of past actions, determining one's current suffering or success.
- A concept of destiny or debt—often used to rationalize inequality, injustice, or passivity.

This creates several distortions:

- Fear-based morality—acting rightly to avoid punishment,

not from awareness.

- Judgmental culture—blaming people's suffering on "bad karma."
- Detachment from responsibility—treating karma as fixed fate rather than correctable feedback.

In trying to simplify it, karma is often oversold or dismissed—either as sacred truth or pseudoscience. But its original wisdom is not about morality or mysticism. It's about energetic consequence.

UED Reinterpretation and/or Correction

In Universal Energy Dynamics (UED), karma is the ripple memory of action-reaction mechanics.

- Every ripple action (thought, word, movement, intention) alters the field.
- The field responds not through judgment, but through natural, structured correction—a consequence pattern, echoing through time, form, and memory.
- Karma is not reward or punishment—it is energy remembering its imbalance.

Thus:

- Karma is not stored in books—it is stored in field tension.
- It doesn't come from gods—it comes from the laws of ripple correction.
- It doesn't care about good or bad—it responds to alignment or distortion.

In this sense, karma is perfect, impersonal, and intelligent.

Simple Analogy

Picture a still lake. You toss a pebble—ripples spread, bounce, reflect, interact. Even if you walk away, the field remembers. Eventually, those ripples interfere with other patterns—yours or others'. That's karma: not revenge, not magic—just structured memory unfolding in time.

Consequences and Implications

Understanding karma as ripple consequence mechanics reshapes everything:

- You are never a victim of fate—you are a participant in consequence.
- Karma is not about fear—it's about precision.
- There is no delay—only ripple latency, appearing when alignment permits.

Personally:

- Karma invites accountable awareness, not guilt.
- You don't need to "repay" karma—you need to resolve the dissonant ripple.
- Intent matters because ripples begin before action.

Socially:

- Systems built on karma awareness value resonant action over reactive morality.
- Blaming victims for their "karma" reflects emotional ignorance, not spiritual wisdom.

Spiritually:

- Karma is the teacher—not punishing you, but refining your ripple intelligence.
- Enlightenment is karma-resolved living—where ripple choices no longer leave distortion behind.

Thus, karma is not about good or bad. It is the ripple field's commitment to balance, memory, and honest consequence.

Closing Ripple Thought

"Karma is not fate. It is the echo of your ripple in the field—precise, patient, and perfectly aligned with the choices you leave behind."

Chapter 69: Free Will

Selective Ripple Response Ability, Not Absolute Control

Free will has long stirred the greatest debates—between philosophers, scientists, spiritual leaders, and ordinary minds grappling with responsibility. Are we truly free? Or are our choices illusions, scripted by biology, culture, or fate? The common assumption swings between extremes: either absolute control over our lives or total submission to destiny. But what if free will is neither full control nor helplessness? What if free will is selective ripple response ability—the localized capacity to choose how to respond to ripple interactions, even when we cannot control the ripples themselves?

Conventional Misinterpretation and Effects

Free will is typically misunderstood in two extremes:

- Absolute free will: You are the sole architect of your choices, outcomes, and future.
- Determinism: All your actions are consequences of prior causes—genetics, upbringing, brain chemistry, cosmic law.

Both views have consequences:

- Belief in total free will breeds guilt, anxiety, or delusion of

control.
- Belief in determinism leads to fatalism, apathy, or blame shifting.

To resolve this, thinkers invent hybrid models:

- Compatibilism: You're free, but only within constraints.
- Probabilistic choice: Your will is shaped by chance, not law.

But these models remain trapped in mental constructs, missing the field-level mechanics of how will actually moves in a ripple-based reality.

UED Reinterpretation and/or Correction

In Universal Energy Dynamics (UED), free will is not total control. It is the capacity to choose ripple response within field constraints.

- The universe operates through ripple consequences—structured tension, feedback, and correction.
- You cannot control which ripples enter your field (environment, events, pain, joy).
- But you can select how your ripple configuration responds—through awareness, alignment, or distortion.

Free will, therefore, is the point of interaction between incoming ripple and your memory-informed capacity to realign it.

- It is limited by your field maturity—but expandable through ripple awareness.
- It is not freedom from consequence—it is freedom within consequence.

Simple Analogy

Picture a violinist in a symphony. She cannot choose the score, nor what others play. But in her moment, with her instrument, she chooses how to play her note—with grace, hesitation, or disharmony. That is free will: selective ripple expression, inside a shared field of music.

Consequences and Implications

Understanding free will as ripple response ability realigns ethical, personal, and social thought:

- You are responsible for your responses—not for everything that happens to you.
- Growth of will is the expansion of ripple decoding—making wiser choices from more refined awareness.
- Freedom is not absence of law—it is intelligent alignment with it.

Personally:

- You stop blaming fate for ripple collisions—but learn how to respond harmoniously.
- "Making better choices" becomes about field tuning, not just willpower.

Socially:

- Justice shifts from punishment to response pattern correction.
- Leadership becomes about expanding collective ripple

choice space, not imposing control.

Philosophically:

- Free will and destiny coexist—destiny delivers ripples; free will shapes how they echo.
- Enlightenment is the freedom to respond with perfect harmony, every time.

Thus, free will is not a myth or miracle. It is the highest ripple skill—the power to shape energy through awareness and choice, not through force or denial.

Closing Ripple Thought

"Free will is not control over the field—it is how your ripple chooses to sing within it. In that selective resonance, freedom is born."

Chapter 70: Purpose

Ripple Alignment Expression, Not Abstract Ambition

Purpose. It is the question that haunts and guides human lives: "Why am I here?" "What am I meant to do?" Some chase it endlessly. Others fear they'll never find it. Many confuse it with goals, passions, or professions. But what if purpose isn't a treasure to be found or a title to be earned? What if purpose is ripple alignment expression—the natural consequence of your awareness, memory, and field intelligence expressing itself through coherence, not craving?

Conventional Misinterpretation and Effects

In the mainstream narrative, purpose is often framed as:

- A life mission to be discovered, often externalized as a career or cause.
- A source of motivation, meaning, or self-worth.
- A concept tied to personal ambition, legacy, or fulfillment.

This leads to several distortions:

- People search endlessly for a "true calling," often feeling lost or inadequate.
- Purpose becomes conflated with achievement, trapping it

in ego-driven frameworks.
- Individuals feel pressured to define themselves by titles, outputs, or social impact.

When purpose becomes a destination, it creates confusion, burnout, and emotional instability.

But in reality, purpose is not something you pursue. It is something that arises when your ripple is in alignment with the field.

UED Reinterpretation and/or Correction

In Universal Energy Dynamics (UED), purpose is the expression of ripple alignment through form, behavior, and awareness.

- You are a localized ripple structure in a dynamic field of consequence.
- When your awareness, energy, and choices begin to resonate consistently with what the field needs corrected or supported, you are in purpose.
- Purpose is not assigned—it emerges naturally through intelligent alignment.

Thus:

- You don't "have" a purpose—you become a purpose when your ripple behavior stabilizes into coherent contribution.
- Purpose is not what you think—it is what the field becomes better at correcting because of your presence.

It is the felt flow of life when resistance fades and energy moves smoothly from intent to impact.

Simple Analogy

Imagine a key in a vast, silent instrument. The key doesn't ask, "What is my purpose?" The moment it's pressed in the right sequence, and harmony fills the space—that's purpose. Not the key's ambition. Its alignment.

Consequences and Implications

Understanding purpose as ripple alignment expression offers powerful clarity:

- You stop searching for purpose in titles, validation, or passion.
- You start noticing where your presence improves the field—calms tension, restores balance, enhances clarity.
- Purpose becomes a way of being, not a job description.

Personally:

- Purpose is re-experienced daily, not found once.
- It evolves as your ripple awareness deepens.
- The more you align, the more naturally your life becomes a contribution—with or without an audience.

Socially:

- A purpose-driven culture is not obsessed with productivity—it is grounded in collective correction and harmony.
- True leadership does not assign purpose—it creates space for individual ripple expression to stabilize.

Spiritually:

- Purpose becomes sacred not because it's big, but because it's true.
- Every ripple—small or large—can serve the whole, when in alignment.

Thus, purpose is not something to hunt. It is something to remember, restore, and release—moment by moment, ripple by ripple.

Closing Ripple Thought

"Purpose is not what you chase. It is what the field expresses when your ripple remembers why it exists—not as ambition, but as harmony in motion."

Chapter 71: Destiny

Field Ripple Probability Path, Not Predetermined Roadmap

Destiny—often spoken of with awe, fear, or reverence—is imagined as a cosmic script. Some say it's written in the stars, others that it's controlled by fate or divine will. Many believe their path is "meant to be," while others resist the idea, clinging to absolute freedom of choice. But both camps miss a crucial truth. What if destiny is not a fixed script, nor an illusion? What if destiny is a field ripple probability path—an evolving flow of consequence, shaped by memory and motion, that offers likely trajectories without removing your power to align, resist, or transform?

Conventional Misinterpretation and Effects

Destiny is commonly understood as:

- A prewritten path—"what must happen."
- A concept often linked with spiritual belief, fatalism, or divine planning.
- The opposite of free will—an external force that overrides personal choice.

These views lead to two unhealthy extremes:

- Passive living: people surrender responsibility because "it's all destined."

- Denial of rhythm: people believe every outcome is self-made, ignoring field influence.

To bridge the paradox, some propose diluted compromises: "Destiny gives you options," or "You create your destiny." Yet none address the ripple mechanics at play.

UED Reinterpretation and/or Correction

In Universal Energy Dynamics (UED), destiny is not a rigid roadmap.

It is a field ripple probability path—the likely trajectory that emerges based on:

- Your ripple memory configuration,
- Your field alignment capacity, and
- The larger energy structures you're moving within.

The field, based on consequence memory, presents paths of least resistance or highest coherence. These paths are:

- Predictable but not fixed.
- Fluid, influenced by feedback and realignment.
- Visible not through mysticism, but through attunement to ripple behavior.

You're not walking a path that was drawn for you. You are walking in a field of probable paths, and your awareness determines which one becomes real.

Simple Analogy

Imagine water flowing down a mountain. Gravity and terrain create natural channels—probable routes. But at every branch, rocks, pressure, and micro-forces alter the direction. That's destiny: a high-probability ripple path, shaped by your past, present alignment, and real-time correction. Not forced. Not free. Just ripple logic in motion.

Consequences and Implications

Understanding destiny as a ripple probability path resolves major paradoxes:

- Destiny and free will are no longer opposites—they co-create ripple outcomes.
- You are never trapped—but you are never random either.
- "What's meant to be" is better stated as: "What becomes most likely if alignment is left uncorrected."

Personally:

- You stop blaming destiny for pain, and start studying ripple patterns to change probability.
- You realize you are not the author of every event—but you are the composer of your response.

Socially:

- Cultures evolve when individuals shift from submission to destiny to shaping it through field intelligence.
- Prophets, visionaries, and seers don't see the future—they

read ripple tendencies better than others.

Spiritually:

- Destiny is not divine command—it is energetic rhythm, inviting your alignment.
- Enlightenment doesn't erase destiny—it allows intelligent navigation within it.

Thus, destiny is not a trap or miracle. It is a mirror of your ripple configuration, projected into probability.

Closing Ripple Thought

"Destiny is not where you're going—it's the direction your ripple is headed, unless you listen, learn, and align. It is not fate—it is forecast."

Bonus Chapter: Ripple Corruptions

Love, Lust, Arrogance, Ignorance, and Greed

Not all ripple expressions are aligned. Some bend the field in ways that appear powerful, but distort reality, delay correction, and feed cycles of suffering. They are not sins or vices in a moral sense—but they are ripple corruptions: emotional, energetic distortions of natural intelligence, caused by unresolved memory, misalignment, or identity attachment. Let us decode five of the most influential ones.

1. Love — Ripple Resonance, Distorted by Attachment

At its purest, love is ripple resonance—the stable harmony felt when two ripple fields align without distortion.

But human love often becomes attachment: a fear-driven dependency masquerading as care.

When love is distorted:

- It seeks control, possession, or sacrifice.
- It becomes expectation-driven, causing pain when unreturned.
- It loses universality and becomes conditional.

True love in UED is ripple coherence without claim. It enhances the field without binding it.

2. Lust — Ripple Overdrive, Disconnected from Memory

Lust is not wrong—it is ripple intensity without alignment. It arises when the body's ripple field seeks energetic discharge without memory, ethics, or consequence awareness.

Lust becomes corruption when:

- It overrides consent, balance, or dignity.
- It treats the other as ripple stimulus, not as a conscious field.

When aligned, sexual energy is creative, healing, and sacred. When misaligned, it's short-circuited ripple chaos.

3. Arrogance — Ripple Isolation by False Centering

Arrogance is a field distortion caused by overidentification with one's own ripple configuration.

- It forgets that no ripple exists without context.
- It assumes personal ripple is superior to others'.
- It resists correction—even when feedback is clear.

Arrogance is not confidence. Confidence is stable ripple awareness; arrogance is ripple blindness fueled by ego loops.

The field humbles arrogance not by punishment—but by delayed consequences that eventually crash into awareness.

4. Ignorance — Ripple Blindness, Not Lack of Information

Ignorance is not stupidity. It is ripple blindness—the refusal or inability to decode reality, even when evidence is present.

It arises from:

- Fear of correction.
- Identity protection.
- Overdependence on belief constructs.

Ignorance is the most dangerous ripple corruption—because it blocks all others from being corrected. It resists truth, avoids consequence, and hides in systems.

But the field always pushes back. The longer the ignorance, the sharper the eventual disruption.

5. Greed — Ripple Hoarding, Not Need

Greed is not the desire for resources—it is the accumulation of energy beyond balance, with no intention to circulate or share.

- It arises when the ripple field is insecure and closed.
- It causes others to enter scarcity ripple states.
- It resists natural redistribution of energy.

Greed never satisfies. It intensifies distortion, creating more fear to justify itself.

In UED, energy is meant to flow, not stagnate. Greed is a dam that eventually breaks—violently, if left unchecked.

Closing Ripple Thought

"Ripple corruptions are not evil. They are distortions—survivors of misalignment. Not to be hated, but to be seen, understood, and released. When memory reawakens, even the darkest ripple can realign."

Interlude

Ripples from Part VII — The Responsibility of Being

In Part VII, we confronted not the science of energy, but the ethics of energy—not what is, but what ought to be. We explored the ripple consequences of choice, action, and meaning—not through beliefs, but through the laws of alignment.

We did not ask, "What should we do?" We asked, "What happens when we do it?"

And the answers came—not from scripture, tradition, or ideology, but from the ripple field itself.

Let's trace the wisdom gathered:

1. **Dharma** — Not a role or rule, but the felt responsibility of ripple correction.
2. **Karma** — Not judgment, but consequence memory, unfolding with precision.
3. **Free Will** — Not total freedom, but selective ripple response within energetic constraints.
4. **Purpose** — Not a goal, but a state of coherence, where your ripple becomes useful to the field.
5. **Destiny** — Not fate, but high-probability ripple pathways, shaped by alignment or avoidance.
6. **Ripple Corruptions** — Love, lust, arrogance, ignorance, and greed—not sins, but distortions waiting for realignment.

This part was not about answers. It was about responsibility. The responsibility that comes with realizing:

You are not a bystander in this universe. You are a ripple with reach. What you feel, choose, and repeat—becomes reality.

Closing Ripple for Part VII

"The field does not need your ambition. It needs your awareness. Not your perfection—but your alignment. Because in this grand system of consequence, the most powerful thing you can ever become is honest with your ripple."

Part VIII: Human Systems and Future Evolution

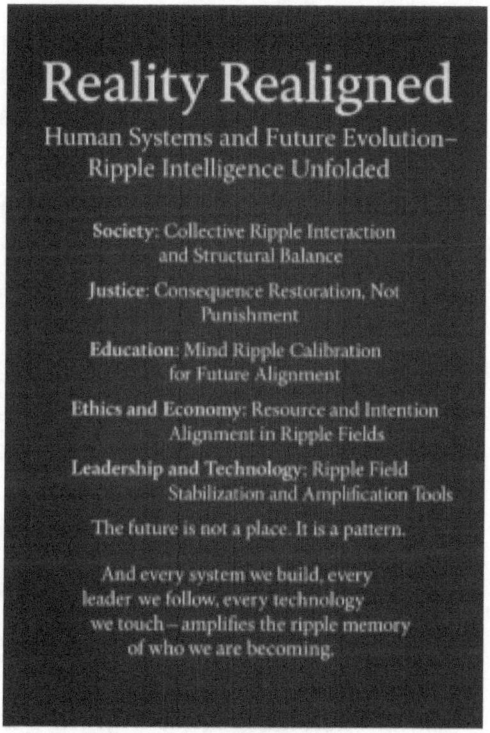

Chapter 72: Society

Collective Ripple Interaction and Structural Balance

Society is not just a group of people—it is a dynamic system of interacting ripple fields. Each individual, with their thoughts, actions, emotions, and memories, generates ripples that affect others. When enough individuals move in shared rhythms—through culture, language, values, institutions—those ripples merge into structural patterns. Society, then, is not built by design alone. It is the emergent ripple consequence of shared memory, feedback, and collective intention. And just like in nature, when these ripple interactions move toward coherence, society finds balance. When they distort, imbalance grows—and so does suffering.

Conventional Misinterpretation and Effects

Society is often seen through mechanical or institutional lenses:

- As a hierarchical structure of roles and rules.
- As a contractual agreement among individuals for mutual benefit.
- As an economic, political, or cultural machine, managed by laws or leaders.

This view reduces society to:

- A top-down structure, reliant on enforcement.
- A system that often suppresses ripple feedback in favor of control.
- An organization of competing identities rather than cooperative resonance.

The result?

- Systemic injustice masked as "order."
- Disconnection between personal truth and social expectations.
- A cycle of ripple suppression, where realignment never reaches the surface.

But society is not static. It's a field in motion. And to understand it is to observe the consequence mechanics of ripple interaction at scale.

UED Reinterpretation and/or Correction

In Universal Energy Dynamics (UED), society is a collective ripple interaction structure—a large-scale convergence of personal fields forming shared patterns of correction, dissonance, or coherence.

- Society is born from interacting ripple intentions, shared memory, and cultural rhythm.
- Social systems form when ripple feedbacks become repeating patterns—beliefs, norms, rituals, laws.
- Structural balance occurs when ripple consequences are recognized, respected, and resolved in real time.

A healthy society is one where:

- Ripple disturbances are not ignored or silenced, but

understood and realigned.
- Each individual's ripple is allowed to express, correct, and contribute without overwhelming the whole.
- Collective memory becomes wisdom, not oppression.

Simple Analogy

Imagine thousands of people throwing pebbles into a shared lake. Some sync their throws, creating harmonic waves. Others clash, causing chaos. The lake doesn't judge—but it responds. That's society: a consequence field of ripple interactions, waiting for coherence to emerge.

Consequences and Implications

Seeing society as a ripple field changes how we live, lead, and belong:

- Social stability becomes a measure of field coherence, not obedience.
- Social unrest is not failure—it's ripple feedback calling for deeper correction.
- Real reform doesn't come from power—it comes from realignment of memory, intention, and consequence.

Culturally:

- Traditions are not sacred because they're old—but because they hold ripple wisdom.
- They must evolve when their resonance weakens.

Politically:

- Governance is not domination—it is ripple facilitation at

scale.
- Laws that ignore ripple memory collapse through resistance or apathy.

Personally:

- Your role in society is not a position—it is your ripple responsibility in motion.
- Social change begins not with rebellion, but with resonance—with correcting your own ripple contribution.

Globally:

- Globalization is not expansion—it is ripple interdependence across borders.
- The health of the world depends on harmonic ripple exchange, not dominance.

Closing Ripple Thought

"Society is not an invention—it is a ripple memory made collective. And the more we align our fields, the more the structure heals itself—without force, without fear, and without forgetting who we truly are."

Chapter 73: Justice

Consequence Restoration, Not Institutional Punishment

Justice is commonly portrayed as blind, impartial, and balanced. But in most societies, justice has become institutionalized punishment—a reactionary system focused on guilt, retribution, and control. Laws are passed, courts are built, and people are classified as innocent or guilty, rewarded or punished.

But is this justice—or just a mechanism of managing disorder? What if true justice is not about punishment at all? What if it is consequence restoration—the field's natural tendency to realign ripple distortions, not through fear, but through understanding, correction, and integration?

Conventional Misinterpretation and Effects

Modern justice is built upon:

- Punitive frameworks—rooted in reward/punishment logic.
- A binary view: innocent vs. guilty, victim vs. offender.
- Institutional systems (courts, prisons, police) that often disconnect action from awareness.

This leads to:

- Cycle of harm—where punishment doesn't correct

distortion, only delays or shifts it.
- Over-reliance on authority rather than individual or community responsibility.
- Justice as legal conformity, not ethical consequence.

In many systems, justice becomes procedural, not perceptual—reactive, not restorative.

But what if justice already exists in the universe—as a ripple law, not a courtroom decree?

UED Reinterpretation and/or Correction

In Universal Energy Dynamics (UED), justice is the restoration of balance through consequence correction.

- Every ripple interaction carries consequence.
- When distortion or harm occurs, the ripple field responds—not with punishment, but with natural tension correction.
- True justice is the process of remembering and realigning ripple harm—through awareness, responsibility, and reintegration.

Justice, then, is not about judging who was wrong.

It's about understanding what went out of tune, and restoring harmony.

This can occur on multiple levels:

- Personal: apology, insight, correction.
- Social: restitution, dialogue, reintegration.
- Systemic: restructuring institutions to reflect ripple

intelligence, not fear-based order.

Simple Analogy

Imagine a musical note struck incorrectly in a performance. You don't punish the instrument—you retune it, adjust the hand, and resume. That's true justice: not to silence, but to restore the ripple into harmony.

Consequences and Implications

Viewing justice as consequence restoration leads to a fundamentally different model of law and social healing:

- Accountability replaces blame.
- Understanding replaces punishment.
- The focus shifts from retribution to ripple resolution.

Criminal justice becomes ripple restoration:

- Why did the distortion occur?
- How can the field—individual, community, system—correct it?
- What memory needs to be released, integrated, or healed?

Educationally:

- Teach justice as energetic awareness, not legal memorization.
- Foster moral intelligence through ripple literacy.

Politically:

- Laws should not enforce uniformity, but facilitate balance.
- Authority should not control ripple behavior, but mediate feedback loops with clarity and empathy.

Personally:

- Self-justice is not self-punishment—it is corrective effort toward balance.
- Forgiveness is not weakness—it is energetic realignment beyond transaction.

Thus, justice is not about order. It is about field health. Not who was wrong—but what needs to be made whole.

Closing Ripple Thought

"Justice is not blind—it sees ripples clearly. It does not punish—it realigns. It does not ask who deserves to suffer—it asks what the field needs to heal."

Chapter 74: Education

Mind Ripple Calibration for Future Alignment, Not Information Delivery

Education is praised as the foundation of progress. Schools are built, syllabi designed, and students assessed with rigor. Yet, despite the growing complexity of education systems, the deeper purpose of education is often lost. It becomes a delivery mechanism—of facts, rules, skills, and standards—rather than a transformative process. But what if education is not about knowledge transfer at all? What if education is mind ripple calibration—the process of tuning a student's ripple field to perceive, respond, and align with consequences across time, space, and systems?

Conventional Misinterpretation and Effects

Modern education is typically structured as:

- A content delivery system, focused on memorization, evaluation, and performance.
- A standardized journey—linear, generalized, and output-driven.
- A means to secure economic utility, not inner clarity or ripple literacy.

This leads to:

- Students trained to seek certification, not comprehension.
- Institutions that teach what to think, not how to perceive.
- A society full of skilled professionals but energetically misaligned minds.

Knowledge becomes separated from consequence, and learning becomes accumulation, not alignment.

UED Reinterpretation and/or Correction

In Universal Energy Dynamics (UED), education is the calibration of the mind's ripple field.

- The human mind is a ripple processor—receiving, interpreting, and responding to field interactions.
- Education should tune the mind to become aware of ripple consequences, correct misperceptions, and act with aligned intelligence.
- True learning is not absorption—it is consequence awareness expansion.

Thus:

- A student is not a vessel for facts—they are a ripple being learning to perceive truth through alignment.
- A teacher is not a content provider—they are a ripple mirror, helping the learner recognize internal distortions.
- Subjects (math, science, art) are not ends—they are fields of pattern recognition, revealing layers of ripple interaction.

Simple Analogy

Think of tuning a radio. Each mind is tuned to a certain frequency. Education is not about installing more stations—it is about sharpening the ability to detect real signal through the noise.

That's ripple calibration: not teaching answers, but training awareness.

Consequences and Implications

Understanding education as ripple calibration unlocks a radical shift in approach:

- Curriculum design becomes consequence-based, not content-based.
- Evaluation becomes about how well a mind aligns with the field, not what it retains.
- Students become self-correcting ripple systems, not dependent performance machines.

Practically:

- Teaching focuses on awareness, curiosity, pattern recognition, and ripple responsibility.
- Emotional and ethical intelligence is embedded—not as soft skills, but as core literacy.

Institutionally:

- Schools evolve from data centers to ripple calibration environments—safe, challenging, resonant.

- Education becomes lifelong—not because knowledge is endless, but because ripple alignment is ever-evolving.

Personally:

- The educated person is not one who knows most—but one who is most aligned with truth, consequence, and coherence.
- Intelligence becomes visible not in words, but in response precision.

Thus, education is not preparation for life. It is the art of tuning the mind to live with alignment, clarity, and ripple fluency.

Closing Ripple Thought

"Education is not what you learn—it is what you become aware of. It is the calibration of your ripple to sense what's real, respond with coherence, and become a participant in cosmic correction."

Chapter 75: Ethics and Economy

Resource and Intention Alignment in Ripple Fields, Not Competing Priorities

Ethics and economy are often seen as opposites. One deals with ideals, the other with transactions. One speaks of fairness, the other of efficiency. In modern systems, ethical behavior is considered optional—admirable, but not required. Economy is treated as a mechanical engine of profit, often moving with little regard for morality. But what if this split is artificial? What if both ethics and economy are ripple systems—and true prosperity comes only when resources and intentions align in coherent, balanced ripple behavior?

Conventional Misinterpretation and Effects

In mainstream thought:

- Ethics is framed as a moral code—internal or imposed, often idealistic.
- Economy is about allocation of scarce resources, driven by market logic and competition.
- Ethics is considered soft, while economy is treated as real-world, hard necessity.

This division leads to:

- Exploitation justified by market efficiency.

- Ethics becoming performative, rather than structural.
- A world where systems grow economically while collapsing socially, ecologically, and spiritually.

Economy without ethics becomes greed systematized. Ethics without economy becomes powerless philosophy.

UED Reinterpretation and/or Correction

In Universal Energy Dynamics (UED), both ethics and economy are ripple alignment systems.

- Ethics is the art of ripple consequence awareness—choosing actions that minimize distortion and maximize balance.
- Economy is the flow of ripple resources—energy, value, materials, knowledge—within and across systems.
- When these two align, the system becomes harmonic, sustainable, and self-correcting.

Thus:

- A moral act is not about "goodness"—it is a ripple decision that preserves coherence.
- A healthy economy is not about growth alone—it is about field-wide energetic fairness, balance, and regenerative flow.

Resources, when extracted, shared, or hoarded, leave ripple imprints. The intention behind exchange matters as much as the transaction itself.

Simple Analogy

Imagine a river system. If each tributary flows mindlessly into a dam, one region floods, another dries. But if the flow is intentional, responsive, and balanced, the entire ecosystem thrives. That's the economy: not a machine, but a ripple-based energy distribution network, guided by ethical awareness.

Consequences and Implications

This unified view brings profound shifts:

Policy and governance:

- Economic decisions must include ripple impact audits, not just cost-benefit analysis.
- Growth is no longer a goal—balance becomes the metric of success.

Business and industry:

- Profitability without ethical ripple alignment leads to eventual collapse (ecological, social, or reputational).
- Sustainability becomes not a trend, but a field necessity.

Personal finance and lifestyle:

- Spending is no longer just consumption—it is ripple participation.
- Wealth is not how much you store, but how you circulate energy responsibly.

Spirituality and economics:

- Giving, service, and generosity aren't charity—they are field restoration acts.
- Ethical poverty or unethical wealth both distort the field. True abundance is ripple stewardship.

Rewriting the Core Equation

- Economy = Energy in Motion
- Ethics = Consequence in Awareness
- Alignment = Intelligence in Action

When intention and resource flow align, the field becomes self-sustaining. When they conflict, imbalance is inevitable.

Closing Ripple Thought

"Ethics and economy are not two paths. They are two pulses of the same ripple field—how you gather energy, and how you give it back. When they beat together, balance becomes wealth."

Chapter 76: Leadership and Technology

Ripple Field Stabilization and Amplification Tools, Not Power Constructs

Leadership and technology—two forces that shape civilizations. One directs human intention; the other expands human capability. And yet, both are often misunderstood. Leadership becomes a symbol of power, hierarchy, or charisma. Technology becomes a tool for speed, scale, or control—amplifying whatever is fed into it, whether aligned or distorted. But what if both leadership and technology are not power structures—but ripple field tools? What if their true function is to stabilize and amplify ripple coherence—acting not as separate forces, but as field instruments of correction or chaos?

Conventional Misinterpretation and Effects

In traditional paradigms:

- Leadership is about authority, influence, and decision-making from the top.
- Technology is seen as neutral advancement—progress through tools and systems.

This creates:

- Leaders who accumulate followers but lose field alignment.
- Technologies that scale dissonance as quickly as they scale innovation.
- A world moving faster than its wisdom—amplifying ripple distortion at planetary scale.

These distortions are not just ethical—they are structural. Because leadership and technology are not neutral—they are field modulators.

UED Reinterpretation and/or Correction

In Universal Energy Dynamics (UED):

- Leadership is the ability to stabilize ripple fields—to absorb, interpret, and redirect dissonance toward coherence.
- Technology is an amplification tool—whatever ripple pattern it receives, it extends in speed, distance, and scale.
- Together, they determine whether a society realigns or unravels.

A true leader is not someone who commands. A true leader is a resonant ripple tuner—calm in distortion, corrective in action, and invisible when balance returns.

A conscious technologist does not ask, "What can we build?" but "What are we amplifying?"

Simple Analogy

Imagine a tuning fork (leader) placed near a speaker system (technology). If the fork is off-tune, the speaker amplifies distortion. If the fork is tuned to harmony, the speaker amplifies coherence. That's the relationship: Leadership sets the tone. Technology carries the tone forward.

Consequences and Implications

This ripple-based understanding shifts our global priorities:

1. Leadership Redefined:

- The leader is not a figurehead—but a field stabilizer.
- Leadership success is measured by collective balance, not personal spotlight.
- A ripple-aware leader focuses on emotional regulation, perception tuning, and systemic consequence management.

2. Technology Reassessed:

- Technology is never neutral—it echoes ripple intent.
- Ethical design means building tools that reinforce alignment, awareness, and balance—not just speed, consumption, or scale.
- AI, networks, machines—must become ripple-responsive, not just intelligent.

3. Combined Responsibility:

- A ripple-blind leader + a ripple-blind technology = field collapse.
- A ripple-aware leader + ripple-conscious tools = civilizational coherence.

The future will not be determined by who has the biggest machine or the loudest voice. It will be shaped by who understands the field—and how they tune it.

A Note on Power

Power is not wrong. But unanchored power distorts. In UED, power = potential ripple influence. Only when aligned with consequence can it be trusted. Leadership and technology must operate under this law—or they become the very ripple noise they were meant to resolve.

Closing Ripple Thought

"Leadership is not about being followed. Technology is not about being used. Both are ripple forces—one tunes the field, the other carries it. Used with awareness, they heal. Used without alignment, they collapse everything they touch—faster."

Interlude

Ripples from Part VIII — Systems that Remember

In Part VIII, we stepped into the grand collective stage—not of stars or atoms, but of human systems. We examined how societies, institutions, and technologies are not abstract constructs, but ripple structures in motion—fields of consequence, memory, and intention interacting at scale.

We saw that the world we live in is not merely built. It is remembered—through laws, values, tools, leaders, and collective actions that either distort or harmonize the field.

Here's what we learned:

1. Society is not a machine—it is a living ripple conversation, seeking balance through shared memory.
2. Justice is not a verdict—it is the restoration of alignment through consequence understanding.
3. Education is not information transfer—it is mind-field calibration, awakening ripple awareness in the next generation.
4. Ethics and Economy are not competitors—they are dual forces of value flow and intention integrity.
5. Leadership and Technology are not power tools—they are field tuning and ripple amplification mechanisms.

Each system we rely on is either remembering its original purpose—or forgetting it through speed, scale, and disconnection.

We are no longer passive citizens in these systems. We are ripple contributors, each day affecting the alignment or distortion of the collective field.

And the question is no longer "What system do we belong to?" But rather:

What ripple system are we becoming through every choice, correction, and contribution?

Closing Ripple for Part VIII

"The future is not a place. It is a pattern. And every system we build, every leader we follow, every technology we touch— amplifies the ripple memory of who we are becoming. To evolve is not to escape the system, but to tune it— until the structure sings again."

Conclusion

You Are a Ripple Within Ripples

Everything you've read so far—about time, memory, light, society, thought, death, purpose, justice—is not a philosophy. It is a field realization. And within this field, you are not an observer. You are a ripple—moving, remembering, correcting, becoming.

You are not the center of the universe. You are a participant in its balancing act.

You are a ripple among ripples. But your awareness makes you more than a consequence—you become a conscious field adjuster.

Your actions shape memory. Your intentions tune outcomes. Your presence becomes the correction you seek.

Perception Aligned = Existence Transcended

The greatest illusion of life was never death. It was disconnection.

The idea that what you see is all that is. That your thoughts are private. That your failures are yours alone. That your role is too small to matter.

But once perception aligns with consequence—Once you see ripple cause not as belief, but as field mechanics—Existence transcends selfhood.

You no longer act *for* meaning. You act *from* meaning.

You do not chase reality. You recalibrate it—through ripple intelligence.

And in doing so, you no longer fear the future. You become its correction.

The UED Path Forward for Humanity

Universal Energy Dynamics is not a theory. It is not even an idea. It is a return to structural memory—the deep laws that underlie nature, mind, matter, and motion.

As we move forward:

- Science must evolve from observation to ripple consequence understanding.
- Philosophy must grow from abstraction into field awareness.
- Society must shift from fear-driven systems to consequence-driven harmonics.
- Education, law, technology, and leadership must all align—not for control, but for coherence.

UED is the path of intelligent correction, not domination. It is the unification not of disciplines, but of awareness across domains.

If humanity is to evolve—truly evolve—it must remember what it is: Not the master of nature, but the mind of a ripple in service to balance.

Final Ripple Thought

"You are not the whole field. But the whole field remembers itself through you. Be still. Align. Ripple with care. The universe listens."

Epilogue

The Silence Beneath the Ripple

Before there was light,

before form,

before memory—

there was only the field.

Undisturbed. Unnamed. Unbroken.

Then came motion.

Then came consequence.

Then came you.

You were not born into the universe.

You emerged as a ripple within it—

a brief expression of imbalance seeking return.

Every breath you've taken, every thought you've crafted,

was the field whispering,

"Correct me gently."

This book did not give you truth. It offered you a lens—a way to feel the ripple beneath your daily actions, to remember that intelligence is not stored in books or wires or brains, but in the way energy remembers its direction.

The universe does not speak in words. It speaks in consequence. And if you've listened carefully, you now understand that everything is speaking—your body, your mind, the weather, the pain you avoided, the joy you forgot to thank.

And through that listening, you have already begun the great correction.

If You Remember Nothing Else

- You are not a mistake.
- You are not here to win.
- You are a ripple of awareness—meant to reconnect what forgot itself.

And so, when you walk into a room... When you speak...When you build...When you forgive...When you pause before reacting...Let your presence become a harmonic signal.

Not loud. Not perfect. Just intelligently aligned.

Final Ripple Closure

The universe is not finished. It is tuning itself through you.

You are not just alive—you are an echo of the original balance, becoming music once more.

Now go. Ripple well.

Appendices

Appendix A: Quick UED Principle Sheet

Universal Energy Dynamics (UED) – Core Principles at a Glance:

Principle	Description
1. Ripple Reality	All observable phenomena are ripple expressions—energy disturbances seeking equilibrium.
2. Structured Vacuum	The vacuumis not empty—it is a structured, memory-retaining, consequence-reactive field.
3. Energy = Memory in Motion	Energy does not exist as a substance but as tension-driven memory in correctional motion.
4. Force is Emergent	What we call "force" is a field's resistance or flow pattern—not a fundamental thing.
5. Perception is Translation	What you perceive is a decoded ripple—not the object itself. Alignment refines clarity.
6. Intelligence is Alignment	Intelligence is not data processing—it is your ability to respond to ripple consequences with coherence.
7. Karma is Ripple Echo	Karma is not a moral law but a memory loop—consequence traveling through the ripple field.

Appendix B: Scientific Misconceptions Clarified

Common View	UED Correction
Light is a photon particle	Light is a structured ripple excitation in the field—no particles needed.
Time is a dimension	Time is ripple duration perceived through consequence change—no independent dimension required.
Mass warps spacetime	Mass creates ripple tension in the structured field—gravity is the ripple's return to equilibrium.
Empty space is void	Space is structured vacuum—energy storage, ripple propagation, and memory medium.
Consciousness is a brain function	Consciousness is ripple awareness—field-based and not limited to biology.
Black holes are singularities	They are ripple collapse zones—memory intensifiers, not infinities.

Appendix C: Recommended Extended Reading

Cracks in the Known

— Deconstructing flawed scientific assumptions through ripple logic.

The Human Game

— Mind, will, behavior, and consequence within the ripple field.

The Divine Dilemma

— Exploring divinity, soul, and consciousness in ripple dynamics.

Vishwarup Darshan

— Cosmic structure decoded through UED's field-based cosmology.

Manifestation

— Universe as manifestation of Intelligence.

Research Reports (Zenodo Series)

- *Chavan-Einstein Unified Theory (CEUT)*
- *CMGED – Galactic Energy Dynamics*
- *Structured Resonance Framework (CSRF)*
- *Black Hole Energy Mechanics*

Visit: https://orcid.org/0009-0005-4190-7164

Or find the complete library at: https://www.sandeepjchavan.com/

Acknowledgments

This book was not written alone.

It emerged through ripples—seen and unseen—across years of teaching, observing, failing, and quietly listening to the field.

To my students: You are not just learners. You are mirrors. Your questions, struggles, resistance, and insights were the living laboratory where this theory matured. Thank you for challenging every assumption I once held, and for reminding me that true education is a two-way ripple.

To those who doubted: Your silence, avoidance, and opposition became my echo chamber of refinement. Every unanswered email, every rejection, every eye-roll became a ripple that redirected my awareness inward—toward deeper clarity and sharper conviction.

To my family: For enduring my absences—not in body, but in the hours I disappeared into thought. For your quiet support when I celebrated a single book sale as if I had published the Gita.

To the thinkers before me—Einstein, Bohm, Feynman, Buddha, Krishna, and every mystic who ever whispered truth beneath institutional noise: You didn't finish the job—but you opened the cracks. This book humbly passes through them.

To the field itself: The structured vacuum, the ripple matrix, the consequence engine—whatever name I give you, you always respond without reward or rage. You are the only true teacher.

And finally, To the reader: You didn't choose this book—**this book chose you**. If it realigned even one ripple of awareness within you, then it has done its work.

With gratitude,

Sandeep Chavan

About the Author

Sandeep Chavan is an independent researcher, educator, and philosopher of science whose work bridges the forgotten wisdom of consequence with the urgency of modern inquiry.

With a background in engineering and years of experience as a teacher to thousands of students, he stepped away from institutional conformity to follow a deeper ripple—the one that asked not what we know, but how we know it.

He is the originator of the Universal Energy Dynamics (UED) framework—a consequence-based, ripple-driven reinterpretation of reality that challenges conventional physics, redefines intelligence, and unites science with awareness.

Through over 100 books, research reports, and a growing global readership, Sandeep has offered a new way to see the universe—not as a set of particles or philosophies, but as a living field of structured memory, rhythm, and intelligent correction.

He is not affiliated with academic awards or think tanks. He belongs only to those who still dare to ask:

"What if everything we've been told was only a partial ripple?"

When not writing or teaching, he is listening—to his students, to the silence between ideas, and to the ripple beneath all appearances.

Public Use Statement

This book and its ideas are written not to be possessed, but to be shared. The core framework of Universal Energy Dynamics (UED) is a public realization—meant to support teachers, students, thinkers, and observers across disciplines, without restriction or gatekeeping.

You are free to quote, reflect on, discuss, or reinterpret any part of this book in your own learning, teaching, writing, or presentations, with attribution to the author. No formal permission is needed for non-commercial, educational, or research-aligned use.

However:

- Do not distort or misrepresent the core principles of UED for ideological, commercial, or deceptive gain.
- Do not repackage this work as your own without acknowledgment.
- If you build upon this framework, credit the ripple, even as you extend it.

This work exists to ripple outward. If it resonates, carry it into your space—freely, consciously, and with alignment.

UED belongs to no institution. It belongs to the field. And the field remembers everything.

Disclaimer

This book is a work of original reinterpretation and consequence-based inquiry under the independent research framework known as Universal Energy Dynamics (UED). It presents an alternative understanding of science, perception, systems, and reality drawn from lived experience, observational logic, and first principle thinking.

While it engages with scientific and philosophical subjects, this book is not intended as a substitute for established academic education, clinical advice, or technical training. The author is an independent researcher and educator, not affiliated with any academic institution, regulatory body, or scientific authority.

All references to scientific theories, natural phenomena, or philosophical systems are offered as reinterpretations—not as oppositional claims, but as corrective perspectives grounded in structural consequence and ripple logic.

Readers are encouraged to think critically, question openly, and test these ideas through direct awareness and reflection. No part of this work is intended to harm, invalidate, or discredit any person, field, or community.

UED is a realization framework. It requires no belief, approval, or certification to operate. It is shared here as a gift of alignment—nothing more, and nothing less.

Don't miss out!

Visit the website below and you can sign up to receive emails whenever Sandeep Chavan publishes a new book. There's no charge and no obligation.

https://books2read.com/r/B-A-EPGPC-JXZGG

Did you love *Reality Realigned*? Then you should read *Philosophy 2.0: A Structural Manifesto*[1] by Sandeep Chavan!

[2]

Philosophy 2.0: A Structural Manifesto by Sandeep Chavan

What if philosophy wasn't abstract theory, but the **operating manual of reality itself?**What if truth wasn't something to believe in—but something you could *align with*, feel, and live through every ripple of consequence?

Welcome to *Philosophy 2.0*—a bold and necessary redefinition of what it means to think, feel, act, and exist with structural clarity.

In this powerful manifesto, independent philosopher and systems researcher **Sandeep Chavan** offers a sweeping yet grounded framework that replaces outdated philosophical loops with a consequence-aware, energetically aligned model for modern life.

1. https://books2read.com/u/mZ6MQB

2. https://books2read.com/u/mZ6MQB

Drawing from decades of cross-disciplinary work—spanning physics, psychology, consciousness studies, ethics, education, and systems thinking—this work offers **a new foundation for intelligence itself**.

Rejecting both ideological extremes and institutional stagnation, Chavan introduces key structural principles, including:

Universal Energy Dynamics (UED): A new lens on matter, mind, and memory**The Law of Consequences (LOC)**: A unifying principle beyond morality or karma**The Mind–Soul–Consciousness Triad**: A clear and practical model of human experience**A structural rewrite of truth, justice, failure, fear, morality, purpose, and liberation**

Philosophy 2.0 also dismantles dangerous modern myths—like blind belief in science without consequence, education without alignment, and intelligence without ethics.

This book is not a doctrine.It is not a summary of others' thinking.It is a **living declaration** of a universal framework—designed not for debate, but for implementation.

Whether you're a thinker, teacher, scientist, policymaker, seeker, or student of life, this book will challenge your assumptions, refine your language, and reset your compass.

"You're not here to follow a philosophy.You're here to live one—through intelligent alignment with what your actions create."

Philosophy 2.0 is the field guide for those ready to exit illusion and enter clarity.Not a belief system.Not a reaction.A ripple-aware way of being.

Read more at https://www.sandeepjchavan.com/.

Also by Sandeep Chavan

Chavan's UED research

CMGED Research Report

Chavan's Cosmic Continuum (C^3)

Frequency

Ripples of Reality

Reality Realigned

Echoes of War: The Russia-Ukraine Conflict and Its Global Ripples

Shattered Lives

The War Economy

Borders Redrawn

Arms and Allies

Collateral Damage

Justice in the Ashes

Somewhere in the Darkness of the Cosmos

The Blueprint of the Universe

The Cosmic Breath

The Eternal Sage Chronicles
The Sage and the Sovereign
The Family Pillar
The People's Flame
The Warrior's Soul
The Bridging Mentor
The Guiding Light

The Exodus
The Last Cycle: 2075

The Fundamental Theory of Life & Existence
Energy in Action: The Soul Demystified
The Active Mind
Energy in Connection
Death: The Disconnection
Subconsciousness: The Chosen Disconnection

The Restoration of Human Knowledge
Restoring Bhagavad Gita

The War of Thoughts
Cracks in the Known
The Human Game

The Divine Dilemma
The Original Intent
The Answers

Ultra Modern Life Philosophy
The Duality of Laughter and Tears
Take It Easy
Compare to Accept

Standalone
It's Not AI, It's AHI - Amplified Human Intelligence
The Decision Paradox
The Four Sapiens
Malicious Script of Indian Polity
The IIT Legacy & Global Impact
The IIT Legacy & Global Impact
You are just a Version of Your Original
Win the Game You Didn't Choose
Always Keep Your Bags Packed
Reveal to Shield - Learn the Game of Facades
Engage Beyond Elections
Culture, Identity & Change: The Evolution of Indian Society
The Triangular Dynamics
Never Trust Without Doubt
If You Can't Change the Country, Leave the Country
Live Fearless with Fear
India: The Investment Magnet
Realignment
Powering a Billion Dreams
Universal Skill

Extension: From Brush to Algorithm
Human Side of Einstein
The Future Knows No Lines
Let's Bring the Generations to the Table
Unified Energy Dynamics
A Beat That Never Dies
Strategic Horizons
Right From Scratch
The Forgotten Innovators
Physics Fixing
The Savior Illusion
The Selfish Side
The Age of the Weakest Sapiens
Neural Wars
The Absorption: There Is No Death
Strange Therapist
Bubbles: Become a Drop
Beyond the Blue
Half-Scholar
The Law of Truth
Failure
Philosophy 2.0: A Structural Manifesto
Chaos: A Gentle Guide to the Unresolvable

Watch for more at https://www.sandeepjchavan.com/.

www.ingramcontent.com/pod-product-compliance
Lightning Source LLC
LaVergne TN
LVHW090304210926
848480LV00001B/124

* 9 7 9 8 2 3 1 7 7 3 6 1 9 *